GUIDELINES FOR FORENSIC ENGINEERING PRACTICE

EDITED BY
Gary L. Lewis

SPONSORED BY
Forensic Engineering Practice Committee
Technical Council on Forensic Engineering of the
American Society of Civil Engineers

Published by the American Society of Civil Engineers

Library of Congress Cataloging-in-Publication Data

Guidelines for forensic engineering practice / prepared by Forensic Engineering Practice
Committee, Technical Council on Forensic Engineering, American Society of Civil
Engineers ; editor-in-chief, Gary L. Lewis.
   p. cm.
   ISBN 0-7844-0688-X
   1. Forensic engineering. I. Lewis, Gary L. II. Technical Council on Forensic
Engineering (American Society of Civil Engineers). Forensic Engineering Practice
Committee.

TA219.G85 2003
620--dc21   2003052211

American Society of Civil Engineers
1801 Alexander Bell Drive
Reston, Virginia, 20191-4400
www.pubs.asce.org

Any statements expressed in these materials are those of the individual authors and do not necessarily represent the views of ASCE, which takes no responsibility for any statement made herein. No reference made in this publication to any specific method, product, process, or service constitutes or implies an endorsement, recommendation, or warranty thereof by ASCE. The materials are for general information only and do not represent a standard of ASCE, nor are they intended as a reference in purchase specifications, contracts, regulations, statutes, or any other legal document. ASCE makes no representation or warranty of any kind, whether express or implied, concerning the accuracy, completeness, suitability, or utility of any information, apparatus, product, or process discussed in this publication, and assumes no liability therefore. This information should not be used without first securing competent advice with respect to its suitability for any general or specific application. Anyone utilizing this information assumes all liability arising from such use, including but not limited to infringement of any patent or patents.

ASCE and American Society of Civil Engineers—Registered in U.S. Patent and Trademark Office.

Photocopies: Authorization to photocopy material for internal or personal use under circumstances not falling within the fair use provisions of the Copyright Act is granted by ASCE to libraries and other users registered with the Copyright Clearance Center (CCC) Transactional Reporting Service, provided that the base fee of $18.00 per article is paid directly to CCC, 222 Rosewood Drive, Danvers, MA 01923. The identification for ASCE Books is 0-7844-0688-X/03/$18.00. Requests for special permission or bulk copying should be addressed to Permissions & Copyright Dept., ASCE.

Cover photos, top, from left: Hatchie River Bridge[1], railroad bridge abutment[2], shattered roof [3], failed beam connection[3]. **Bottom, from left**: failed beam connection[3], flood damage[4], Schoharie River Bridge[5].

From [1] Wiss, Janney, Elstner Associates, Inc.; [2] Union Pacific Railroad, Structures Design Department; [3] Donald Kraft, P.E.; [4] Kenneth M. Meikle, Nebraska Department of Water Resources; [5] U. S. Federal Highway Administration, National Highway Institute

Preface

Every noble work is at first impossible.

— Carlyle

In its 1985 charter the to Forensic Engineering Practice Committee, the Technical Council on Forensic Engineering (TCFE) of the American Society of Civil Engineers (ASCE) tasked the committee with creating this set of *Guidelines for Forensic Engineering Practice*. Work on this document was initiated in 1994 and completed in 2002. This eight-year development period exemplifies the difficulty and magnitude of effort in composing such a document. The guidelines have undergone numerous internal and peer reviews, generally resulting in impassioned exchanges regarding the content and style of the text. After innumerable informal reviews and revisions, final formal review comments were provided by TCFE's Executive Committee on August 21, 2002, resulting in consensus by all involved.

Within the broad field of Civil Engineering, the practice of Forensic Engineering involves the investigation of performance difficulties of buildings, pipelines, structures and other facilities. Investigation of failures usually involves an interface with the legal system, most often in the form of expert testimony. The overall purpose of the set of guidelines is to commit to writing the current state of Forensic Engineering Practice. As such, this document will not be a standard or code.

The set of guidelines is organized into general areas of interest. They are *Qualifications, Investigations, Ethics, Legal Forum and Business Considerations*. An introduction chapter sets the stage for the need, timeliness, overall purpose and general organization. A hypothetical failure scenario is presented in Appendix A for illustrative purposes. The intent of the hypothetical is to demonstrate the complexities of failures and create a basis from which the context of points in each chapter can be demonstrated, especially by those unfamiliar with concepts of Forensic Engineering.

The *Qualifications* chapter addresses commonly accepted education and experience requirements for Forensic engineers. The term "expert" is discussed as defined both by the courts and by the engineering profession. Various aspects of Federal and State law will be cited as they apply to the Engineers offering expert testimony. Admissibility and disqualification are mentioned in this chapter and further expanded in subsequent chapters.

The *Investigations* chapter is intended to show the typical aspects of physically carrying out a forensic investigation. This chapter is not meant to train the engineer to perform an acceptable investigation. The thrust is to demonstrate what constitutes appropriate practice. The handling of evidence for subsequent courtroom presentation is addressed. The proper use of building code standards in investigation and analysis is described. The logical progression from investigation to final opinion is put forth. Other issues such as alternative opinions are addressed.

The *Ethics* chapter is the primary focus of the guidelines. Promulgation of guidelines for ethical behavior of the Forensic engineer is a primary objective of ASCE and the Forensic Practice Committee. The ASCE Code of Ethics and Codes from other engineering associations are applied to the Forensic engineer. Advocacy, objectivity, conflicts of interest, and the appearance of such are defined and discussed in detail along with numerous other ethics topics. The ethical aspects of expert opinions, fees and time constraints are put forth. Means of reporting unethical practices and other sanctioning processes by the regulatory bodies and ASCE are presented.

The *Legal Forum* chapter gives a brief overview of the court system as it applies to the construction industry. The role of the Forensic engineer as an expert witness is defined. The interactions of the Forensic engineer with plaintiffs, defendants, attorneys for each side and other witnesses, both experts and factuals, are addressed. The role of the Forensic engineer in non-adjudication forums is discussed.

The *Business Considerations* chapter is intended to relate the non-technical management side of Forensic Engineering practices. Marketing of Forensic Engineering services within an acceptable ethical scheme is encouraged. The need for contracts, insurance and liability is addressed. The issue of compensation for services as it relates to both the Legal Forum and Ethics is discussed.

The authors wish to express their sincere thanks to all of the exceptional professionals who have given selflessly over an extended period of time toward this product, in particular, Lewis L. Zickel, who served as Senior Technical Reviewer on behalf of the TCFE Executive Committee. Appreciation is expressed for the support of the TCFE Executive Committee as well as the ASCE staff.

Forensic Engineering Practice Committee

Gary L. Lewis, Ph.D., P.E.	Theodore G. Padgett, P.E.
Member	Past Chairman

Acknowledgments

Forensic Engineering Practice Committee

Keith E. Brandau, P.E., S.E., Chairman
Theodore T. Padgett, P.E., Past Chairman
Rubin M. Zallen, P.E.
Harvey A. Kagan, P.E.
Kim Beasley, P.E.
Gary L. Lewis, Ph.D., P.E.
Robert L. Johnson, P.E., S.E.
Leonard J. Morse-Fortier, Ph.D., S.E.
Robert T. Ratay, Ph.D., P.E.
Anthony Dolhon, P.E.

Contributing Companies

Frauenhoffer and Associates
Parsons
Packer Engineering
Hudson International
Wright Padgett & Associates
Simpson Gumpertz & Herger, Inc.

Sponsors

Technical Council on Forensic Engineering of the
American Society of Civil Engineers

Contributing Authors

Chapter 1: Introduction
Theodore G. Padgett, Jr.,Wright, Padgett & Associates, Inc., 1941 Savage Rd, Ste 400C, Charleston,
SC 29407

Chapter 2: Qualifications
Gary L. Lewis, Parsons, 1700 Broadway, Ste 900, Denver, CO 80290
Steven Zebich, Packer Services Corporation, Packer Engineering, P.O. Box 353, Naperville, IL
60566-0353

Chapter 3: Investigations
Robert L. Johnson, R.L. Johnson & Associates, Ltd., 525 Crescent Blvd, Glen Ellyn, IL 60137
Leonard J. Morse-Fortier, Simpson Gumpertz & Herger, Inc., 297 Broadway, Arlington, MA 02474-
5310

Chapter 4: Ethics
Gary L. Lewis, Parsons, 1700 Broadway, Ste 900, Denver, CO 80290
Keith E. Brandau, Frauenhoffer and Associates, P.C., 3002 Crossing Ct, Champaign, IL 61822

Chapter 5: Legal Forum
Theodore G. Padgett, Jr. Wright, Padgett & Associates, Inc., 1941 Savage Rd, Ste 400C, Charleston, SC 29407
Gary L. Lewis, Parsons, 1700 Broadway, Ste 900, Denver, CO 80290

Chapter 6: Business Aspects
Harvey A. Kagan, Hudson International, 565 E. Swedesford Rd, Wayne, PA 19087
George R. Barbour, Kimley-Horn & Associates, P.O. Box 88068, Raleigh, NC 27636

Manuscript Peer Reviewers
Rubin M. Zallen, Zallen Engineering, 1101 Worcester Rd, Framington, MA 01701
Ronald E. Martell, Moore, Costello & Hart, Ste 1350, Craig Hallum Center, 701 Fourth Avenue South, Minneapolis, MN 55415-1823.
Lewis L. Zickel, Consulting Engineer, 93 Myrtle Ave, Dobbs Ferry, NY 10522.
Ozzie Rendon-Herrero, Mississippi State University, P.O. Box 1172, Oxford, MS 39762
Harry Teeter, Advanced Land Engineering, Inc., P.O. Box 0916, Estero, FL 33928-0916
Steven Zebich, Packer Engineering, Inc., 1950 N. Washington St, P.O. Box 353, Naperville, IL 60566-0353
John A. D'Onofrio, D'Onofrio Engineers, PC, 172 South Ridgedale Avenue, East Hanover, NJ 07936
Robert T. Ratay, Ratay Inc., 198 Rockwood Road, Manhasset, NY 11030

Contents

CHAPTER 1 – INTRODUCTION

The Forensic Practices Committee shall enhance the competent and ethical practice of Forensic Engineering by developing practice guidelines and seminars and conference sessions aimed at elevating the understanding of professionals engaged in the investigative and Judicial arenas.

- Charter, ASCE Committee on Forensic Practices, 1985

1.1 PURPOSE OF GUIDELINES

Engineering, Architecture and Contracting are all businesses that function by making a profit. Some percentages of the design and construction work performed in this country are flawed. Most defects are non-consequential while others cause major calamities.

Determination of the cause of a performance deficiency in a building, pipeline, structure or other engineered facility is the purview of the Forensic engineer. The findings of the Forensic engineer are usually presented in the form of expert opinions within the legal system. The need for Forensic Engineers to remain unbiased and carry out their duties in an ethical fashion is a major theme of this set of guidelines.

The relationship between the legal profession and the engineers who work within it is quite clear. Attorneys are advocates and engineers are not. A lawyer's function is to shape his or her client's position into the legal stance that is most likely to prevail in court. The engineer's role is to present expert opinions fairly and objectively without regard to the outcome of the legal dispute. The engineer must be an advocate only of his or her opinion.

It is the intent of this set of guidelines to simply describe the current state of Forensic Engineering practice and provide the guidelines mentioned in our committee charter. Accordingly, chapters on Qualifications, Investigations, Ethics, the Legal Forum and Business Considerations are presented.

1.2 HYPOTHETICAL CASE STUDY

A hypothetical case, including many of the complexities of actual cases, is included in Appendix A for illustrative purposes. Each chapter concludes by discussing the applicability of the chapter content to the hypothetical. In order to garner the maximum benefit, the reader should familiarize himself or herself with the hypothetical before delving into the text. The hypothetical itself is totally fictitious in nature. Any perceived similarity to a past failure is purely accidental.

The hypothetical selected deals with a bridge failure but just as easily could be a building or other Civil Engineering project. The facts surrounding the failure, the parties to the ensuing dispute and findings of the various technical entities are given in the Appendix.

1

CHAPTER 2 - QUALIFICATIONS OF FORENSIC ENGINEERS

It is not good to have zeal without knowledge.
- Proverbs 19:2

2.1 INTRODUCTION

As society becomes more complex and as technological advances occur, litigation involving highly complicated topics becomes commonplace. The need has skyrocketed for experts with specialized technical knowledge who can skillfully explain their knowledge and provide relevant opinions. Experts play a significant role in investigating failures and presenting their findings in court. In addition, plaintiffs, defendants, attorneys, judges, and juries are being asked more and more to believe and rely upon opinions of the experts, a phenomenon known as "experto credite."

2.1.1 Chapter Content

What makes any Forensic engineer an "expert?" This chapter summarizes the current understanding of what qualifies an engineer for rendering expert opinions and testimony. Because the engineering profession and the courts have different criteria for establishing who qualifies as an expert, both are presented in this chapter. As with material in other chapters, qualifications described in this chapter are intended as a guideline rather than a standard for establishing or measuring qualifications of Forensic Engineers. After summarizing these criteria, the chapter relates the criteria to the hypothetical presented in Appendix A. The chapter concludes by providing a checklist for testing individual qualifications of Forensic Engineers.

No university offers an "expert engineering" or a "Forensic Engineering" curriculum, so the state of being an expert doesn't come from a degree. Recognition as an expert Forensic engineer by the profession or the courts requires more than a formal college education in the subject matter. Conversely, an engineer having 10, 20, 30, or even 50 years or experience does not necessarily imply attainment of "expert" status. Whether by education or experience (or both), being qualified as an expert requires a level of achievement and mastery of the topic that allows the engineer to form accurate opinions and deduce correct conclusions. This chapter provides descriptions of this level of achievement and mastery from both the profession's and the court's perspectives.

2.1.2 Attributes of Expert Engineers

Both the engineering profession and courts accept the following attributes as helpful in qualifying an engineer for this type of work:

- An undergraduate degree in engineering, preferably from a program accredited by the Accreditation Board of Engineering and Technology/Engineering Accreditation Commission (ABET/EAC)

2

- Technical competence in the subject area – defined by the engineer's *own* professional work in studies, design, research, teaching or writings.
- Graduate school degrees in the subject area.
- Professional license, obtained by examination, to practice in the subject area.
- Full time practice in the relevant field of expertise.
- Authorship of peer-reviewed publications on the subject, preferably recently.
- Authorship of textbooks on the subject.
- Active involvement with professional organizations that serve to advance the relevant technical subject.
- Extensive, similar experience on other projects involving the same subject.
- Awards or peer recognition for accomplishments in the subject area.
- Objectivity, honesty, relevance, thoroughness, and professional demeanor.
- Good citizenship.
- Previous occasions of being accepted by other courts as an expert.
- Confidence in his or her own qualifications and opinions.

Stated another way (which has legal significance as shown later); the key attributes of an expert engineer are *education, training, experience, skill and knowledge.* Any engineer possessing these attributes, measured by possessing any number of the bulleted items above, and who can perform his or her work *accurately, objectively* and *in a professional manner* would most likely qualify and capably serve as an expert in his or her field.

2.2 QUALIFICATIONS OF FORENSIC ENGINEERS

2.2.1 Forensic Engineer Qualifications – Two Views

Black's Law Dictionary defines an expert as, "one who is knowledgeable in a specialized field, that knowledge being obtained from either education or personal experience." The same dictionary defines an expert witness as, "one who by reason of education or specialized experience possesses superior knowledge respecting a subject about which persons having no particular training are incapable of forming an accurate opinion or deducing correct conclusions." Note that the definition of an expert witness used in the legal profession, such as the one from Black, does not compare the knowledge or experience of the engineer with peers, but instead compares it to *the knowledge of laypersons.* This chapter elaborates on this issue and presents qualifications of experts from both the engineering and legal perspectives.

The definition of an expert recommended by the Forensic Practices Committee as a guideline for the engineering profession is, "any individual whose knowledge, skill, education, training, professional experience, absence of bias and peer recognition indicate superior knowledge about a particular field of endeavor such that the foundation exists to provide factual and authoritative conclusions and opinions." Though this definition doesn't require recognition by peers, peers are best equipped to determine whether an engineer meets this metric. As shown below, this definition encompasses the credentials that are generally agreed upon by both the courts and the engineering profession. Some differences exist, such as the court's acceptance of laypersons rather than peers as the determinant of whether one is an expert. It is the position of ASCE that the engineering profession should define what qualifications an engineering expert should have and that the profession is better able than any others to judge whether an engineer meets the criteria (ASCE 1977).

Normal procedure followed by courts in qualifying and admitting experts for courtroom testimony is to first establish that the subject matter of the testimony is in an area in which the trier of fact will benefit by the assistance of someone with expertise in the subject matter. Next, the court must be convinced that the particular engineer proposed for testimony has the training, skill, and experience to provide the assistance.

No textbook or court document provides a convenient checklist of the qualifications for experts. Based on reviews of numerous cases, an engineer having any subset of the attributes listed above would likely satisfy most courts and the engineering profession that the person has the expertise for giving authoritative, "expert" opinions. Though qualifications as viewed by the court and by the engineering profession are described here, it should be noted that the final determination in and legal forum is made by the individual court.

Qualification as an "expert" comes to engineers from two camps. The first, in the order presented in this chapter, comes from the legal profession. Because the definitions are different, engineers may fail to be recognized by their profession as an expert but might meet the test of the legal profession. As shown in this chapter, legal definitions and tests of qualifications of experts are more important to the jurisprudence system. This set of guidelines maintains that a Forensic engineer should meet both the legal and technical criteria. Thus, an expert could be defined as one who meets both tests of qualifications.

The engineering industry recognizes, as do the courts, that for an individual to be accepted as possessing sufficient *expertise and mastery,* he or she must demonstrate sufficient *education, training, experience, skill,* and *knowledge.* These are referred hereafter as *Key Attributes.* It is asserted in this set of guidelines that the profession, as well as the court, should prescribe which attributes define acceptable technical qualifications for providing expert testimony. It is also acknowledged, as shown below, that the ultimate determination in jurisprudence rests with the court.

2.2.2 Qualifications of Forensic Engineers as Viewed by the Courts

Though an engineer may have achieved his or her profession's respect as an expert, the *court* rather than the individual's profession decides whether the person qualifies to testify as an expert in that jurisdiction. It is important to point out that *voir dire* from opposing counsel can only be directed to the five key attributes. Therefore, it is essential that the engineer be prepared to provide sufficient information on these attributes when first taking the stand to convince the judge or jury that he or she is truly an expert authority on the subjects for which testimony will be provided.

Federal rules of evidence have been relaxed and their latitude widened in regard to who can qualify in court as an expert and what testimony is admitted. Under Federal Rule of Evidence 702[1] and other similar state and local statutes and cases, a witness need only be qualified "by knowledge, skill, experience, training, or education." Consideration of including the adjective "special" with these requirements was given in early drafts of the rule but was considered too restrictive. In fact, case law has indicated that an expert need not have complete knowledge of his or her field, or even be certain of his or her opinions.

2.2.2.1 Admissibility

Considerable debate exists over admissibility of testimony by experts. Admissibility is discussed in detail in Chapter 5, but is mentioned here because of the implication that an important qualification of Forensic Engineers, from the court's perspective, is that their testimony be admissible.

Until recently, it was unclear whether the criterion for admissibility in *Daubert versus Merrell Dow* (see Chapter 5) applied to testimony by engineers. The Supreme Court ruled in that case that trial judges could assess admissibility of expert testimony on the basis of four criteria:

- Peer review and publication,
- Rate of error,
- Empirical testability, and
- General acceptance within the scientific community.

In March 1999 the Supreme Court, in *Kumho Tire Co. Ltd., et al., versus Patrick Charmichael, et al.*, settled this issue by ruling that judges can apply the same four tests previously targeted at medical and scientific professionals to expert testimony by engineers. This ruling is significant not only in the legal forum (Chapter 5), but also in regard to qualifications (Chapter 2) and ethics (Chapter 4). An engineer's qualifications may be challenged if his or her opinion differs with standards of practice or is not consistent with generally accepted methods, especially if cross-examination reveals a lack of understanding of accepted practices. Many opinions rendered by engineers are based on theories or techniques that are not compatible with

---

[1] Rule 702 of the Federal Rules of Evidence state that testimony must first of all assist the trier of fact, "If scientific, technical or other specialized knowledge will assist the trier of fact to understand the evidence or to determine a fact in issue, a witness qualified as an expert by knowledge, skill, experience, training, or education may testify thereto in the form of an opinion or otherwise."

the norms of the discipline. The ruling becomes relevant in the ethics arena if the witness testifies falsely about the standards or that the standards of practice aren't applicable (these and other breaches are discussed in detail in Chapter 4). It is allowable to testify that the standards were not prudently followed, but it is unethical to distort the standards or argue that a new, not generally accepted, method should have been used. While admissibility of testimony is testable, the reliability of testimony comes down to personal ethics.

2.2.3 Qualifications of Forensic Engineers as Viewed by the Engineering Profession

The general view in the engineering profession is that any licensed engineer should be able to provide testimony that would benefit the court in understanding technical matters. It is also generally held that practicing Forensic Engineers, and the courts, do not view all licensed engineers as experts. However, the profession unilaterally accepts licensing as a minimum requirement for practice.

Forensic Engineers, in general, view licensing as one of the necessary but not sufficient qualifications for expert engineers. Due to this elemental linkage between licensing and qualifying as an expert, licensing requirements are summarized here.

The National Society of Professional Engineers (NSPE 1997) conducted and published a comprehensive survey of requirements for licensure as engineers. The survey summarizes findings reported by all 50 states and 5 territories. The National Council of Examiners for Engineering and Surveying (NCEES) also developed a "model licensing law" for use by states that are formulating or amending their regulations. The NSPE survey included the NCEES model law as another sampling point in the survey.

The NCEES model law recommends, as a minimum, that an applicant for licensure as professional engineer meet the following qualifications:

- Possession of a bachelor of engineering degree from an engineering institution approved by the respective state board,

- Successfully passing two national examinations,

- Demonstrating a level of acceptable engineering experience,

- Be of good character and reputation, and

- Provide five references, three of which are from P.E.'s having personal knowledge of the applicant's experience.

NSPE has taken a stronger policy stand on these. They recommend that the NCEES qualifications be met, but add the following requirements:

- That the applicant has a bachelor's degree in engineering (NSPE's position is that a bachelor's degree in engineering technology is *not* acceptable).

- That the bachelors degree be from a program accredited by the ABET/EAC or one assessed by ABET/EAC as substantially comparable,

- That the applicant pass both the national Fundamentals of Engineering and Principles of Practice examinations as prepared and administered by NCEES,

- That the applicant obtain at least 4 years of professional experience after the degree, with experience-credit allowed for graduate study or teaching, and

- That an applicant with both a bachelors and Ph.D. from an ABET/EAC accredited institution, or any engineering faculty member with a degree from an ABET/EAC accredited institution, be excused from taking the Fundamentals of Engineering examination.

The NSPE survey found that all except six states require a minimum of 4 years experience, and all but the same six require the applicant to successfully pass a 16-hour examination, normally split evenly between separate fundamentals and principles of practice examinations. Experience is normally checked to assure that it was under supervision of other engineers.

With regard to state rules and regulations that may be related to qualification as Forensic Engineers, the NSPE report reveals the following:

- Only 56 percent of the states require peer recognition, in the form or written references by 3 or more licensed engineers. As noted earlier, peer recognition is one of the fundamental requirements of expert engineers.

- Forty-one states license generically, that is, the engineer may practice outside his or her discipline "to the extent they're qualified."

- A total of 28 states allow licensure without a degree. Generally, these jurisdictions require 10 or more years (20 years in some cases) experience.

- Only 2 states restrict licensure to holders of BS degrees from ABET/EAC-accredited institutions.

- Fifteen states permit licensure to holders of bachelors of engineering technology degrees.

- Twenty-nine states allow licensure to holders of non-engineering bachelor's degrees.

- Only 58 percent of the states require good reputation and character.

- Only 2 states require residency in the state.

- Only 17 states require licensees to pass an examination on state laws and rules.

Of incidental interest, some states may take disciplinary action against engineers for certain activities that might disqualify them as expert engineers. The NSPE survey revealed that:

- Twenty-nine states can take disciplinary action for failure to pay child support,

- Eight states can take disciplinary action for failure to repay student loans,

- Six states can take disciplinary action for failure to pay taxes, and

- Eight states can take disciplinary action for conviction of a felony.

Just as an attorney is qualified to practice law after passing the bar exam, it is generally believed that an engineer is allowed to practice engineering, including some aspects of Forensic Engineering, after achieving professional engineer (P.E.) status. The following paragraphs present several metrics currently used within the profession to recognize and distinguish "expert" engineers.

2.2.3.1 Assistance to Trier of Fact

The first qualification of any expert engineer, both from the professional and legal side, is that the court must first need input from the technical area represented by the engineer (see discussion of Rule 702 in Chapter 5). With the rigor of most engineering college curricula, this broad qualification encompasses practically all graduate engineers.

The engineering profession recognizes that individuals possess varying degrees of expertise. Some among the profession have earned a high level of recognition from others due to their knowledge of very complex issues. Other, less-experienced engineers may also have a high level of expertise, but that expertise may lay several levels of complexity below other, more recognized colleagues. Both types of individuals can provide technical value to court proceedings.

2.2.3.2 Specialty of Engineer's Firm

Another loose measure of an engineer's qualifications is whether he or she has gained and held employment with a firm specializing in the technical field at issue. One of the best means to measure the pulse of the engineering profession, with regard to determining what is considered acceptable, is to scan the *Help Wanted* sections found in the media. Often times, engineering companies are recognized within the industry as being a primary resource for a particular service, such as bridge design, drainage engineering, or digital control systems. Firms such as these may require specific expertise and experience within their respective market. In other instances, the nature of a firm's work may be vast enough to require that an engineer employee possess a broad range of capabilities.

2.2.3.3 Specialty Registration

Many states allow professional registration in specialty areas of engineering. The NSPE survey of all 50 state registration regulations revealed that the following areas are recognized as specialties in various combinations of states (NSPE 1997):

- Civil engineering,

- Chemical engineering,

- Electrical engineering,

- Mechanical engineering,

- Agricultural engineering,

- Structural engineering,

- Mining engineering,

- Geotechnical engineering,

- Petroleum engineering,

- Ceramic engineering,

- Industrial engineering,

- Metallurgical engineering,

- Nuclear engineering,

- Sanitary engineering,

- Corrosion engineering,

- Manufacturing engineering,

- Computer engineering, and

- Fire protection engineering.

It should be noted that most registration designations do not designate any specialty, but instead assign the title, "Professional Engineer." A college curriculum for many of these specialties is identical for the first two years. Note that Forensic Engineering is not in the list, nor is it recognized as a specialty in any of the states or territories polled. These specialty registration areas normally require demonstrated experience in the specialty, as well as in the generic category in which the engineer obtained his or her degree. For example, the engineer may be required to pass the generic Principles and Practices examination, followed by an examination in the specialty area. In general, the profession of engineering has become quite specialized, and is self-regulating by preventing overly broad assertions of expertise.

Regarding Forensic Engineering qualification issues, a dichotomy exists between whom the profession views as a "legitimate" expert or specialist versus a registered engineer who may not be considered to be an expert. Debate arises from whether or not and individual qualifies as an expert based solely on his or her knowledge of the general principles and standards of the industry of a particular discipline, or whether or not the individual has, in fact, participated in performing engineering work directly related to the issue. A P.E. who has four years of work as an E.I.T. and has successfully passed both of the NCEES examinations may have knowledge of the industry standards, but may not have worked in the corresponding discipline. The profession has not reached a consensus on this issue. Some engineers believe that one must have physically *designed* a particular facility to be able to "opine" on an issue related to it. Others believe that if an individual possesses sufficient *knowledge, skill, experience, training and education,* he or she then possesses sufficient technical expertise to assist the "trier of fact" and originate opinions on the issue.

The debate continues over specialty registration, and will for some time. Recent developments in professional associations reveal a trend toward smaller, highly specialized groupings of engineers. It is sufficient to say that within the arena of Forensic Engineering, failure investigation involves many issues that cannot easily be dissected, or made black and white. For example, an attorney once inquired about finding an expert on strings. When advised that perhaps there may not be a *"string expert"* at large, and that perhaps a specialist trained in microscopy and hemp manufacturing technology may be of assistance, the attorney said "thank you", and kept searching for a string expert.

2.2.3.4 General Registration (Professional Engineer)

Another more definitive measure of the qualifications of a candidate "expert" engineer is whether the engineer has been licensed to practice engineering in one or more states. Licensing in most states requires the candidate to spend 4 or more years as an engineer in training and pass written (and often oral) examinations in the relevant field of practice and code of conduct. After passing the initial examination, most states, through reciprocity or comity, will accept a candidate who is licensed in another state by means of a statement of experience and payment of a fee.

Many states also require proof of professional development or continuing education as a requirement for renewal of licenses. State requirements for professional engineer registration and renewal can be found by contacting the state board of registration for professional engineers, or by referring to the NSPE survey of all state regulations (NSPE 1997).

Regardless of the country, region or state, the engineering profession, by and large, recognizes that certain generally accepted principals, practices and experience contribute to create acceptable credentials. Statutes require that before being licensed, engineers must possess minimum levels of education and apprentice-type training under the direction of an experienced registered professional. Upon completing the necessary training, the engineer must then demonstrate a minimum level of competency, usually via examination. Upon successful completion of the legally mandated minimum requirements, the title of *Professional Engineer* (P.E.) is bestowed.

An engineer's professional practice must then follow a course that is guided, primarily, by a combination of statutory requirements and the dictates of the profession. It is the dictates of the profession that, more often than not, determine the future acceptance by the profession into the forensic arena as an expert.

Both professional organizations and courts mutually recognize that registration and licensing as a professional engineer may not qualify the engineer as an *expert*. Licensing does require that the engineer has had to pass written examinations to prove his or her ability in the art and practice of work, and licensing binds the engineer to a code of ethics for conduct within and outside the court in accordance with ethical and professional principles.

An important question is whether a license to practice engineering should be viewed as the single and final determinant in assessing whether an engineer is qualified to testify and give expert opinions in courts. Without a doubt, the license and all of its prerequisites qualify any engineer to provide foundation testimony, explaining various aspects of the technology in which they were licensed. General consensus, however, is that the average licensed engineer is not qualified to conceptualize opinions and give *expert* testimony without demonstrating achievement and mastery of the subject beyond the minimum requirements for licensing.

As any engineer who has practiced for a number of years can attest, the gratifying news that one has successfully attained registration (and the title that goes with it), is considered merely indicative of a *new beginning* in professional responsibilities. The levels of responsibility that the *profession* bestows upon the new P.E. are typically dependent solely upon the engineers' demonstrated ability to perform certain tasks with acceptable *accuracy, efficiency* and *timeliness*. These are key attributes of any engineer, but the latter two do not necessarily qualify one as an expert. There are, in fact, many in the profession who have worked within their field for many years without attaining registration. Others may not be efficient or timely. Often times, these individuals have developed a high level of expertise in one or more areas of engineering analysis, design and management. Thus, within the engineering profession at large, expertise is recognized under criteria that go well beyond the attainment of registration. Expertise is plainly and simply substantial knowledge of a particular subject area, with said knowledge being gained by a combination of education and experience.

Finally, the question arises whether an engineer must be registered in the state where testimony is to be given. Either the court or the registration board in any given state may mandate this, and any licensed engineer planning to work in another state should ascertain the requirements. For states that do not mandate registration, it would be prudent for the engineer to become registered if testimony will involve local or state regulations because registration normally requires that the engineer be familiar with relevant state regulations on design. The NCEES has taken the position that an engineer should be registered in the jurisdiction where the testimony will take place. Some states, Nebraska for example, require only that the engineer does not falsely claim to be licensed to practice by the state. Registration in the state is traditionally acquired by engineers, but not considered mandatory in general.

2.2.3.5 Generalist Forensic Engineer

Many failures require that investigations be conducted by more than one specialist. Because clients or counsel are not equipped to mobilize and coordinate investigations by several engineering disciplines, some cases involve the services of a generalist Forensic Engineer. This person serves as the lead investigator who advises the client on the types of specialists needed for the team, selects them with the client's authorization, and coordinates their activities. The generalist may get involved in testimony if qualified to testify in any of the disciplines, or may simply observe the proceedings after coordinating the investigations and assisting with trial preparations.

2.2.3.6 Qualification by Possessing Key Attributes

Another professional (and legal) measure of the qualifications of an expert Forensic engineer is whether the engineer possesses the education, training, experience, and skill to testify and form opinions in their field of endeavor. The key factors most recognized within the engineering community for establishing acceptable technical credentials are associated with an engineer's ability to perform his or her work *accurately, efficiently and in a timely fashion.* These factors provide a useful measure for determining the level an engineer's capabilities. The previously mentioned five attributes may be tested using these factors:

1. Education

Education is the foundation of the engineering profession. Undergraduate work and a bachelor's degree in engineering (preferably from an ABET/EAC-accredited program) provide the fundamental tools with which to perform basic analysis and design within acceptable parameters of *accuracy.* Graduate studies help to refine those basic skills by providing the aspiring engineer with more-advanced techniques and skills that will enhance the *efficiency* of his or her work. A sound theoretical background developed by a good education aids the engineer in selecting the most suitable analytical techniques, thus improving the *timeliness* of the work.

Therefore, when the engineering profession seeks entry-level candidates, the specifics of an individual's education will be scrutinized against how effectively that education facilitates the three factors. Further, the particulars of an individual's education may, in fact, provide a recognized expertise (usually at the doctoral level) which may *uniquely qualify* an individual to perform specific work *accurately, efficiently* and in a *timely fashion.*

2. Training

Training provides the backbone of experience. Education, in most instances, cannot provide sufficient detail to cover all aspects of knowledge within a specific discipline. Most training that an engineer receives is through instruction from an experienced mentor or from more formal training classes. The purpose of technical training is to provide exposure to the particulars of a service area. For instance, the design of buildings involves a wide variety of tasks, from initial conceptual layout, program analysis, code research and preliminary design to material testing and the final detailing of connections and other components. Training then, provides the conduit through which an engineer's education finds practical application. Expertise is developed in one or more areas with time. *Efficiency* and *timeliness* cannot be obtained without the background gained with training.

3. Experience

Experience is perhaps the number one criteria with which the engineering profession uses to measure expertise. It is within the area of experience that *experts* are gauged and scrutinized. To the layman, experience is often viewed as time and time is often viewed as experience. However, the engineering community typically establishes experience levels based on how effectively an individual satisfies the three factors. The profession may deem an individual to be experienced if they can meet the following tests:

a. Has the individual worked within a particular industry long enough to have, under competent direction or under his or her own charge, attained sufficient knowledge and capability to perform their work *accurately?*

b. Has the individual personally designed a particular item a sufficient number of times to have thorough knowledge of configurations, details and variations to the extent that his or her work is *efficient* and *timely* when compared with other members of the profession with similar time in performing such work?

c. Does the individual possess sufficient knowledge, and has he or she demonstrated such knowledge as to be technically competent to perform or direct work, which fully encompasses all appropriate aspects of the work? Some individuals may be perfectly capable of designing a component of a landing gear, but may not possess sufficient knowledge or expertise to design an entire aircraft for example.

4. Skill

Skill is an attribute that is most readily established by how an engineer is recognized by his or her peers. That recognition is a direct reflection of an individual's ability to produce, through his or her own efforts or direction, a product that maintains appropriate technical *accuracy* or suitability. It is ultimately the engineer's duty to the public to not only be accurate, but also to produce that product *efficiently* and in a *timely fashion* if the work is to be successful from an economic standpoint. Thus, the profession at large recognizes that skill is a reflection on how well an engineer satisfies the three factors as evidenced by the level of responsibility the engineer assumes.

5. *Knowledge*

Perhaps more than skill, it is knowledge that defines the term *expert,* for it is within an engineer's knowledge that the culmination of the four previous attributes come to fruition. The attainment, as well as the maintenance of knowledge comes not only from past and on-going experience, but, in the engineering profession, it comes from professional development and activity. Many states now require continuing education for licensed professionals of many career fields. Engineering firms commonly encourage participation in professional organizations as not only a means to develop and increase an engineer's knowledge, but as a means to establish benchmarks within the profession for recognizing technical expertise, professional behavior and qualification.

Knowledge, in the engineering profession, is the reflection of an individual's ability to develop *accurate,* safe and *efficient* solutions to mankind's practical problems. The economics which determine the survival of a firm dictate that they be performed in a *timely fashion.*

Clearly, the engineering profession recognizes that *education, training, experience, skill* and *knowledge* must be reflective of more than merely that which exceeds the knowledge of the *"common man"* or layman. A *qualified* engineer must meet the profession's demands for *accuracy* (in the area in question) to be recognized by his peers as a qualified forensic expert. Evidence of the added traits of *efficiency* and *timeliness* reinforces the probability that the engineer has indeed attained expert status. Registration, in and of itself, does not necessarily reflect the level of an individual's technical expertise.

2.2.3.7 NAFE Qualifications

Another possible measure of peer recognition as a Forensic engineer is senior membership in the National Academy of Forensic Engineers. NAFE is an independent membership organization affiliated with the National Society of Professional Engineers (NSPE). Membership requires adherence to the NSPE code of Ethics, and the association is evolving ethical guidelines specific to Forensic Engineering. Upper membership grades require extensive experience with Forensic Engineering. A nonmember affiliation titled "Correspondent" is available to those interested in the subject, but not qualified for membership. The association address is NAFE, 174 Brady Avenue, Hawthorne, NY 10532, Ph. (914) 741-0623, Fax (914) 747-2988.

2.2.3.8 Professional Society Activities and Senior Membership Grades

Most professional societies and organizations recognize achievement through succession through various grades of membership. ASCE, for example, recognizes lifelong achievement of engineers with honorary memberships or "Fellow" status. These grade levels are not automatic, and require recognition by peers as well as elaborate proof of experience and contributions in their field. Certainly, memberships in societies and positions held on various committees and boards add to the overall and specific qualifications of any engineer for "expert" status.

2.2.3.9 ICED Recommended Practices in Construction Industry Disputes

The document entitled, "Recommended Practices for Design Professionals Engaged as Experts in the Resolution of Construction Industry Disputes" was developed by the Interprofessional Council on Environmental Design (ICED) in 1988. A Forensic engineer may be asked if his or her affiliate professional organizations subscribes to the guidelines. Among others, the NAFE and NSPE have embraced and formally endorsed these guidelines for expert testimony by engineers. The document is silent regarding what qualifies an engineer as an expert, but certain implied qualifications can be gleaned from some of the guidelines presented. For example, the manual states, "Experts should report their need for qualified assistance when the matters at issue call for expertise they do not possess." This implies that one qualification is having sufficient technical mastery of the subject matter to recognize one's technical shortcomings. Another guideline, namely, "the expert witness should testify about professional standards of care only with knowledge of those standards," obviously implies current knowledge of design standards as a qualification for expert status. The document is available through the ASFE, 8811 Colesville Road, Suite G106, Silver Spring, MD 20910.

2.2.3.10 ICED Recommended Practices for Expert Testimony

The ICED recently developed a more relevant document entitled, "Recommended Practices for Design Professionals and Scientists Engaged as Experts for the Technical Review of Others' Work and Providing Testimony in Public Forums" (ICED 1996). As indicated by the title, it addresses both the qualifications and behavior of engineers engaged in providing testimony, and acceptance for membership in any of the endorsing associations might imply that the member meets the qualifications for expert testimony described in the document. As of June 1998, ICED, NSPE, and a few other professional organizations had endorsed the document. As of this writing, ASCE has not endorsed the recommendations. Many of the recommendations address ethical issues regarding demeanor, conducting investigations, conflicts of interest, and other topics, which are described in Chapter 4. The document's recommendations regarding qualifications for expert testimony include:

- Experts should serve only when qualified, by education and experience, in their own areas of competence.

- Experts should be as thoroughly informed as can be considered reasonable through communications and cooperation, including communications with experts retained by the opposing side whenever appropriate and possible.

Copies of the document can be obtained from ICED or NAFE.

2.2.3.11 Other Professional Organizations

Numerous other professional organizations of scientists, medical doctors, police, and others have developed legal definitions, behavior requirements, and presentation procedures for expert testimony. One in particular, the American Academy of Forensic Sciences (AAFS), has an active engineering component. While active membership in any of these may not qualify an engineer for testimony, knowledge of their standards and guidelines would be beneficial. Through its Inter/Intra Professional Liaison Committee, NAFE actively seeks information exchanges with other professional societies and organizations.

2.3 APPLICATION TO THE HYPOTHETICAL

In reviewing the hypothetical presented in Appendix A, it is apparent that more than one qualified, technical expert would be needed to address all the issues. Because two of the three potential litigants cite erosion and scour as operative in the failure, a qualified hydraulic engineer who specializes in sediment transport hydraulics and bridge scour would be needed. Other obvious witnesses needed would include a qualified expert geotechnical engineer, an expert bridge design engineer, an expert bridge foundation engineer, and possibly an expert geologist or concrete specialist.

The river flow rates used by Design Engineer for design were obtained from a government agency but weren't apparently checked or independently evaluated. A qualified expert hydrologist would be needed to address this issue. Many expert hydraulic engineers would qualify for both the scour and hydrologic testimony roles.

2.4 HOW TO ASSESS QUALIFICATIONS OF A CANDIDATE EXPERT WITNESS

2.4.1 Recommended Checklist

For engineers attempting to ascertain their candidacy for expert witness work, and for clients or attorneys seeking assistance from expert engineers, the following fairly-exhaustive checklist of attributes is offered in determining whether the engineer might meet the profession's and court's tests for qualification as an expert witness:

- Does the engineer hold ABET/EAC-accredited college or university undergraduate or graduate degrees in the subject area?

- Does the engineer have technical competence in the subject area – defined by the engineer's *own* professional work in design, research, teaching or writings?

- Is the engineer licensed as a professional engineer and was the license obtained by examination in the subject area,

- Is the engineer's registration in the state in which the failure occurred?

- Does the engineer practice full time in the relevant field of expertise?

- Has the engineer authored peer-reviewed publications on the subject, preferably in the past four years?

- Is the engineer an author of textbooks on the subject?

- Is the engineer involved with professional organizations whose aim is to advance the technical subject?

- Does the engineer have extensive, similar experience on other projects involving the same subject?

- What awards or peer recognition does the engineer have for accomplishments in the subject area?

- Has the engineer demonstrated his or her objectivity, honesty, relevance, thoroughness, professional demeanor, and citizenship?

- Has the engineer had previous occasions of being accepted by other courts as an expert?

- Does the engineer have confidence in his or her own qualifications and opinions?

- Does the engineer have good communications skills?

- Does the engineer have reasonable understanding of legal proceedings and vocabulary?

Many engineers possess most or all of these 15 attributes. These engineers meet the general requirements and will probably masterfully serve the client's and court's need for technical assistance in fact-finding and communicating well-founded opinions to the trier of facts.

2.5 CONCLUSION

Both the engineering and legal professions recognize the common attributes of *education, training, experience, skill and knowledge* as qualifications for expert witnesses. Though peer concurrence that an engineer qualifies as an expert is not required by the courts, the engineering profession alone can judge this metric. Objectivity, discussed at length in Chapter 4, is not listed as a qualification in this common definition, but is an essential ingredient in the conduct of the investigation and proceedings. In summary, superior knowledge about a particular field of endeavor and an absence of bias are required in order to qualify an engineer to provide objective, factual and authoritative conclusions and opinions.

The founding president of the NAFE once stated that the requisite qualification for Forensic Engineering practice is "integrity, competence and hard work…" (Specter 1988). This definition nicely summarizes the qualifications presented in this chapter and is recommended as an initial metric to any client, engineer, litigant, or attorney making initial inquiries into the expertise of a forensic-engineering candidate.

2.6 REFERENCES

ANSI/ASTM undated. *Standard Practice for Reporting Opinions of Technical Experts,* No. E 620-77.

ASCE 1989. *Guidelines for Failure Investigations,* Technical Council on Forensic Engineering, New York, NY.

ASFE 1987. *EXPERT: A Guide to Forensic Engineering and Service as an Expert Witness,* Assoc. of Soil and foundation Engineers, Silver Springs, MD.

ICED 1988. *Recommended Practices for Design Professionals Engaged as Experts in the Resolution of Construction Industry Disputes,* Silver Spring, MD.

ICED 1996. *Recommended Practices for Design Professionals and Scientists Engaged as Experts for the Technical Review of Others' Work and Providing Testimony in Public Forums,* Silver Spring, MD.

Carper, K. L. 1990. "Ethical Considerations for the Forensic Engineer Serving as an Expert Witness," *Business and Professional Ethics Journal,* Vol. 9, Nos. 1 and 2, Spring-Summer.

Carper, K. L., Ed. 1989. *Forensic Engineering,* Elsevier Science Publishing Co., Inc., New York, NY.

Carper, K. L., ed. 1986. *Forensic Engineering: Learning from Failures,* ASCE, New York, NY.

Dolan, T. J. 1973. "So, You are Going to Testify as an Expert," *ASTM Standardization News,* ASTM, West Conshohocken, PA., March.

Feld. J., and K.L. Carper 1997. *Construction Failure 2nd Ed.,* John Wiley, New York, NY.

Friedlander, M. C. 1989. "FORUM: The Design Professions: Let's Regulate Expert Witnesses," *Civil Engineering,* No. 4, April.

Hough, J. E. 1981. "The Engineer as Expert Witness," *Civil Engineering,* December.

Lewis, G.L. 1997. "Objectivity vs. Advocacy in Forensic Engineering," *Proceedings of the First Forensic Engineering Congress,* ASCE, Minneapolis, MN, October.

Murillo, J.A. 1987. "The Scourge of Scour," *Civil Engineering,* ASCE, No. 7.

NSPE 1985. *Guidelines for the P.E. as a Forensic Engineer,* Publication No. 1944, Alexandria, VA.

NSPE 1997. *NSPE Analysis of Professional Engineer Licensure Laws,* Alexandria, VA.

Priedlander, Mark C. 1989. "The Design Professions: Let's Regulate Expert Witnesses," *Civil Engineering,* ASCE, No. 4.

Specter, Marvin, M. 1988. "What Does it Take to be a 'Good' Expert Witness?," *ASTM Standardization News*, ASTM, West Conshohocken, PA, Feb.

TCFE 1989. *Guidelines for Failure Investigation*, Task Committee, Technical Council on Forensic Engineering, ASCE, New York, N.Y.

Veitch, T.H. no date. *The Consultant's Guide to Litigation Services: How to be an Expert Witness*, John Wiley & Sons, New York, NY.

Worrall, Douglas G. 1984. "Engineer Experts – The Attorney's Viewpoint," *Journal of the National Academy of Forensic Engineers*, October.

CHAPTER 3 - INVESTIGATIONS

It is the Glory of God to conceal a matter; to search out a matter is the glory of Kings.

- Proverbs 25:2

3.1 INTRODUCTION

Expert opinions should not be developed, and expert testimony should not be given without first performing an adequate investigation of the failure. This chapter presents an outline of steps that will most likely help assure that a complete investigation is performed. This process will include field investigations, office analyses, and reporting the results of the investigation. It may also include laboratory work or in-situ testing.

3.1.1 Chapter Purpose

No set of guidelines can advise engineers of all the investigations needed for every failure. This chapter does not prescribe a rigid formula for conducting investigations, but instead outlines the fundamental elements of a logical forensic investigation, progressing from initial contact to developing final opinions. These steps are considered essential to an acceptable, professional practice.

3.2 FIELD INVESTIGATIONS

The purpose of the field work in any forensic investigation is to provide the actual visual access to and hands-on observations of an accident, collapse, partial failure, or the as-built conditions of the structure or process under review.

Careful and thorough field investigations provide the data and evidence to assist the office and laboratory investigations. These steps are necessary and sufficient in leading to the development of the professional opinions. The field investigation takes the study beyond the realm of scientific theory into a factual presentation, which can be presented in an understandable fashion. A layperson to the engineering field; whether contractor, insurance adjuster, attorney, mediator, judge or juror; can deal more easily with "show and tell" than with scientific jargon, but many cases may require both.

3.2.1 Identification of Parties Involved

Initial discussions with the client should identify all parties involved in each project. A conflict of interest check should be made prior to the retention of the Forensic Engineer. Ethical considerations in regard to conflicts of interest, interactions with parties, and conducting investigations are provided in Chapter 4. Introductions among those present at any site should take place prior to the day's start-up. The client's attorney will provide the criteria for interaction with the other parties. Ethical considerations in regard to initial discussions are provided in Chapter 4.

3.2.2 Planning and Scheduling

Whenever possible, available documents should be gathered and reviewed prior to the field investigation phase.. The documents will assist in the planning and scheduling effort, and will provide direction for data gathering and equipment requirements. The sections below are presented as a guide to preparing for and conducting the field investigation. Occasionally, a catastrophic failure may call for immediate response. In these cases, the investigator may be called to the site on a moments notice. A camera, field notebook and experience may substitute for more extensive planning in these cases.

3.2.2.1 Staff Selection and Availability

A professional Forensic Engineering office staff may include a variety of team members including principals, licensed engineers, field assistants, and designers. Each field investigation may require a mix of these team members based on the field work requirements. Less-experienced members should be exposed to forensic training, including workshops and seminars conducted by experienced professionals and qualified manufacturer's representatives, classroom work complementary to the training, and hands-on field experience under the guidance of senior staff.

Complex and even simple investigations may require the addition of outside consultants or personnel. Outside team members can provide specialized expertise, as well as guidance to various aspects of the investigation.

The principal considerations in the selection of the field investigation personnel are the project requirements for professional expertise, staff experience and available time, all with consideration of the budgetary limitations of the client. The investigation cannot be sacrificed for budget. Costs can be controlled best if the most experienced and costly staff members work cost-effectively to coordinate and supervise the efforts of the field and office assistants gathering documents, samples, and field data and performing the analyses.

Large forensic firms may employ multiple teams working on individual projects and studies. The small firm may have one person or a small team that works in whole or in part on multiple projects.

3.2.2.2 Equipment Selection and Availability

Each forensic firm will develop an inventory of useful equipment for their normal project requirements. This inventory will tend to grow as the project base expands and as the team experience with data gathering improves.

In some instances equipment and trained personnel may come from other consulting firms or testing laboratories. For example, an experienced equipment operator could be critical to the outcome of the testing and the value of the results, and not all firms employ such operators.

Typical equipment used for projects can include still cameras, video recorders, metal detectors, X-ray equipment, hand and power tools, moisture and chemical detectors, temperature and pressure monitors, and surveying equipment, to name a few. The selection of the essential equipment for a particular investigation should be made by experienced personnel, familiar with the scope of the investigation.

3.2.2.3 Scheduling

Scheduling of the field investigation is dependent on site conditions, staff and equipment availability, and the time and travel requirements of the specific client and study.

Field inspection requirements vary by project. For example, a traffic accident investigation may require accident reconstruction but may be limited by the requirement to reopen the accident scene to normal usage or traffic. In a partial failure, immediate debris removal may be required to allow the reconstruction process to begin. In a failure involving injury or loss of life, rescue and recovery efforts naturally take precedent over investigations. An investigator on site may be called upon to lend special expertise in assisting such efforts. In all cases the investigator should be attentive to the investigation but respectful of rescue and recovery efforts. Many projects are investigated while the facilities in question are in use. This requires special scheduling with the occupants to minimize disruption of their normal routine.

Investigators must be especially mindful of the time and budgetary requirements of the client at the start and during field investigations. Time and budget are often uncertain at the beginning of a project and access to the study site may be controlled by other parties. A preliminary proposal agreement for the initial services gives the client input to fees, services, schedule, and timing. This proposal should consider the added costs of travel, overnight accommodations, and other away-from-office expenses. Continued communication with the client, or their representatives, is essential to the ongoing work progress and the timely collection of fees. When the client requests added services, provide a supplemental proposal with recommendations. Often, during the earliest stages of an investigation, it is helpful to organize the work and budget in phases, with frequent communications between investigator and client.

3.2.3 Mobilization for Site Inspections

The initial site visit can be done by the senior staff member in charge, or by other team members, to evaluate the likely scope of the investigation. If they are not available before hand, overall photos, sketches, and available documents can be collected in the first inspection. This will aid in the formulation of an efficient and cost effective use of team personnel as inspections progress.

3.2.4 Data Gathering and Documentation

The process of gathering information useful for the analysis of a particular project can take on a life of its own. The investigation of a traffic accident allows only a few hours for data gathering with all the failure scene intact, while the investigation of building distress due to soils settlements or expansion can take several years. The sampling and testing for the traffic accident nearly always take place after the evidence is moved and stored at a location away from the accident scene, while building and soils issues stay at the site, but may be changing continually. The following sections provide a general overview of typical forensic fieldwork at a site.

3.2.4.1 Sketches and Notes of Project, Site Conditions

Sketches and notes should be developed to use in conjunction with photographs and/or video. Sketches to scale are useful for later review and discussions with other consultants, but are not always practical under field conditions. Rough sketches with notes included can be redrawn to scale in the office. Rough sketches with photos of the sketch area can be reproduced easily reproduced.

Notes of comments by others, overall field conditions and surrounding terrain and environment will be useful to the development of opinions.

3.2.4.2 Photographs and Video

Film is a useful tool in documenting field conditions, measurements, and test results. Still photography can be used to document patterns, notes, measurements, damages, and scale when objects of known size are included. Site photos should always include a few distant shots, establishing the overall layout of a site, however large. These photos will help team members to visualize an entire site and promote better understanding of the investigation and its context.

Motion photography or video with audio has the advantage of detailed description of the field observation, discussion with witnesses or persons familiar with the past and/or recent history of a project, structure, distress, or collapse.

With the advent of digital photography and the speed of electronic data transfer, many investigators are transitioning away from film photography. Despite the many clear advantages afforded by these new technologies, storing and preserving images poses some special challenges. Compact digital disks offer storage for numerous photographs, but these new media may not provide the robustness and longevity of film negatives. Especially for legal work, the investigator should confer with the client and their counsel before abandoning traditional film photography.

3.2.4.3 Equipment Usage and Calibration

Equipment must be well maintained and calibrated after normal use. The manufacturer's documents and user's manual for each piece of equipment should be kept for reference, and may become exhibits.

Records should be kept for equipment requiring periodic calibration. Most equipment can be calibrated by staff members using manufacturers' guides, while more sensitive tools may require calibration by the manufacturer, or a firm specializing in the sales and service of that particular equipment.

Calibration and checks for accuracy can be determined by measuring known quantities. Example: A metal detector, which measures location and depth of concrete reinforcing, can be verified for accuracy by chipping away the concrete to expose the steel. As a reminder, equipment employed to collect important data should be used only by personnel trained in its use.

Equipment subjected to impact, lengthy time periods between use, or corrosive and dusty environments should be inspected, cleaned, and calibrated prior to use in the field.

3.2.4.4 Evidence Gathering/Chain of Custody

Field samples should be collected with careful documentation. As appropriate, the samples or evidence may be combined with photographs and measurements to confirm their location at the time they were removed.

Samples or test material should be taken under the observation or supervision of the forensic engineer. The forensic engineer may retain these samples in storage or for distribution. Samples for destructive or chemical types of analysis can be given to the approved testing laboratory. Evidence required for other information, such as size or actual damages, may ideally be kept by the engineer. Counsel for all parties to a dispute may establish a formal protocol for evidence gathering, warehousing, and chain of custody. In these cases, the investigator must follow this protocol.

Chain of custody of evidence, to satisfy the court and to withstand vigorous cross-examination, must include full documentation of when, where and by whom the evidence was obtained, who had access to it, and who and where it was stored until its arrival in the courtroom.

3.2.4.5 Sampling and Testing Methods

Forensic sampling is dependent on project quantities. The following examples are given:

- Sampling for the failure of a single steel or wood beam might include a failed member and a similar intact member.

- Studies of distressed structures might require the sampling of as-built construction in 20% of the units in a small development, while only 10% in a large development.

- Investigation of a project constructed in phases, by different contractors or subcontractors, will require sampling in each phase for varying or similar defects.

3.2.4.6 Interviews

Another form of evidence is gathered through interviews or questionnaires. This provides some eyewitness data unavailable through other sources. In multi-structure developments or multiple witness projects, the questionnaire can provide information useful to the study and patterns of repetitive damages or occurrences. Recording interviews is strongly recommended, but nearly always requires permission from the party being interviewed.

Whenever feasible, interviews should be conducted as soon after the event being investigated as possible, while the memories of witnesses are fresh and before lawyers and employers limit or prohibit access to them.

3.2.4.7 Interaction With Media

Discussions or interviews with media representatives are very sensitive in the field of forensic studies. Many forensic projects involve a legal dispute. Published comments by consultants may or may not represent accurately the statements made or opinions offered. These published accounts can be used at a later time to discredit the consultant. Media interaction should be left to the client, governmental bodies, or attorneys, if any. Generally, the forensic engineer should direct to others any and all inquiries by phone or in person from radio, television, or newspaper reporters. No information should be given, unless requested and approved concurrently by the client and their attorney where appropriate.

3.2.4.7 Safety

Failure sites should be considered dangerous and all personnel deployed to investigate such a site must be familiar with basic site safety, and should be equipped and trained to deal with the hazards they may face. Safety glasses, a hardhat, protective shoes and appropriate clothing, hazmat suits, and a respirator may all be required. Where a hazmat suit or respirator are required, their wearer should be fully trained in their safe and proper use.

3.2.5 Data Assembly

The organization and protection of the data collected is critical to the analysis, the office investigation, and when legal action occurs, to the mediation and trial process. This phase of the study is outlined below.

3.2.5.1 Organization and Format of Data

Field data can be organized by numerous methods. The key to this organization is simplicity. As these data may be reviewed by many different persons from various backgrounds and disciplines, its organization should be accessible and understandable by all.

3.2.5.2 Avoidance of Preconceived Notions

Forensic studies and field investigations should gather and record data as found at the site. Every effort should be made to eliminate trying to force the evidence, data, and analysis to fit the consultant's or client's preconceived hypotheses. Do not approach data gathering with an eye toward "how can I find evidence to support my sense that…?" but rather "how can I find all of the important evidence?" When the forensic engineer has been hired by one party to a dispute, one excellent test for bias involves asking "how would I do this if I were working for a different party, or for the public at large?"

3.2.5.3 Storage of Samples and Protection of Evidence

Samples should be stored in a method that maintains them in a condition closest to the way they were found. Ferrous metals should be protected from additional rusting, failed surfaces from rusting, oxidation, impact, crushing, and abrasions. Larger sample storage should provide support for the sample to maintain the shape or configuration existing at the time of field sampling.

3.2.6   Issues Raised in Hypothetical

The following field investigation procedures would have been relevant in the hypothetical bridge failure given in Appendix A:

- The contractor should check components of the bridge structure for conformity to the alternate design drawings.

- The components should be documented through photos, videos, sketches, and measurements. Pier and girder connections at the alternate design changes should be documented.

- Samples of structural steel, concrete, reinforcing steel, prestressing strands, and bolts should be taken for testing by an approved laboratory.

- The laboratory should perform field tests of welds at girders and connectors.

- Core samples at concrete breaks should be tested for aging to attempt to define effects of August 1995 seismic event and June 1996 dam break event.

- Reports and documentation for the seismic and dam breaks, construction inspections, by the State DOT personnel, contractor logs, laboratory tests, deputy inspector and soils reports should be reviewed to help direct the field survey.

- Consultants involved in the field investigation would include civil, hydraulic and structural engineers, a metallurgist, geotechnical and weld certification laboratories, and a general contractor familiar with bridge construction. Some consultants may be retained later, as the study develops.

- In this case, no interaction with the media should occur, unless specifically requested by the client.

3.3 LABORATORY INVESTIGATION

3.3.1 Testing and Inspection

Laboratory investigations normally follow the field investigations. An acceptable laboratory investigations requires selection, performance and documentation of applicable test methods, use of engineering standards, adherence to standard laboratory reporting, and attention to chain of custody. Occasionally, parties to a dispute will agree to a single set of tests with all interested parties invited to observe these tests.

Once the samples are brought to the laboratory, the specimens are tested. These tests must follow approved and accepted procedures, often those established by the American Society for Testing and Materials (ASTM). If there is not a standard procedure, one must determine if the testing procedure is consistent with previous testing programs investigating similar properties. Once the testing is complete, a report can be generated. Where the test standard mandates one, the report must follow the predetermined format. Also, one must be careful to adhere to the agreed-upon chain of custody for the samples throughout the laboratory testing.

3.3.2 Laboratory Reports

Laboratory reports normally contain full documentation of the investigation in each of the following categories:

3.3.2.1 Initial Information

- Testing Laboratory

- Testing Technician and all observers

- Date Tested

- Time Tested

- Test Number

3.3.2.2 Test Information

- Test Type

- Test Standard Followed

- Length of Test

- Test Conditions

3.3.2.3 Testing Equipment Information

- Equipment type

- Calibration Record

3.3.2.4 Sample/Specimen Information

- Sample/Specimen Description

- Sample/Specimen Dimensions

- Sample/Specimen Weight

- Sample/Specimen Storage Prior to Testing

- Sample/Specimen Sketch

- Sample/Specimen Photo

- Sample/Specimen Alterations/Additions (if any)

- Date Obtained

- Sample/Specimen Number

- Sample/Specimen Chain of Custody (described)

3.3.2.5 Test Results

- Test Result

- Plots/Graphs/Curves

- Test Analysis Followed

3.4 OFFICE INVESTIGATION

The axiom, "you can never have too much information", best describes the goal to assemble all available documents and other information when researching the history or origin of a failure or potential failure. When gathering documents and information in addition to the original design drawings and construction-era documents, the investigator should assemble and organize documents and verbal history through closure of the investigation.

Tracking down sources and available documents is often exhilarating as well as frustrating. Most failures occur long after the original principals are no longer involved with the Owner, or the Owner has archived or destroyed many documents describing the design, construction, and history of the project. This makes the search more time consuming and exhausting. However, once elusive and important documents are found and assembled, the investigator can move forward with greater self-assurance. For many, the hunt for and discovery of historical documents is an exhilarating and gratifying experience.

3.4.1 Data Gathering Phase

Data gathering commences with the first notes taken during a telephone call or visit from the client. Even though there may be many principalities involved, the current Owner should be the focal point and origination of the investigation. If capable, the Owner's engineering staff should conduct the initial research and data gathering since the Owner should have design drawing documents contained in the project files. As a minimum, the Owner should assemble the layout drawings describing the structure and its location. From this beginning, the Investigator can re-construct and visualize the design and history of the original project. The Investigator must be creative in his or her thoughts and be freely receptive to all ideas and concepts presented without preconceived notions. There will be many roadblocks (access to Owner's files, uncooperative subjects, etc.) during the data gathering that may hamper the investigation, but an attainable goal and drive to reach that goal will usually prevail.

3.4.1.1 Construction Documents

Starting with a single design drawing (or other form of document information), the investigator may ascertain or discover other leads to follow. The goal is to assemble a comprehensive package of information about the subject of the structure. Categories include:

a. Original Drawings and Calculations - While original calculations represent the "Holy Grail" of the Investigator, it is more probable that design drawings may be the only documentation remaining and uncovered during an investigation. Even so, design drawings often lead to many other referenced items that will aid the Investigator. In the title block alone, the Architect/Engineer (A/E) (who designed the project) should be mentioned with possibly an address and telephone number listed. Initials of names or names along with the drawing signature dates should be given on all title blocks, but the most important identifier is often the project number(s). The project number is generally a unique identifier assigned to a project by whether the Owner and by the designers and their consultants. Project numbers can lead to additional files contained in the Owner's or A/E's document storage that will aid in the investigation. Also, contained on many drawings are reference drawings, design criteria, and other notes, which further expand the Investigator's search and data gathering. Calculations can enable the Investigator to match the analyzed structure with the design drawings and "as-builts" while presenting design criteria, design methodology, and individual structural analyses so that all analyses do not have to be recreated.

b. Geotechnical Reports - For many sites, foundation investigations are conducted prior to beginning design work. Foundation investigations are commissioned by the Owner and copies are usually available in the Owner's files. In addition to subsurface explorations and foundation recommendations, site specific seismic time histories or acceleration coefficients may be included in a geo-technical report. As a minimum, the geotechnical report includes boring logs (soil classification) and boring locations of the existing site soils. Recommended soil bearing capacity, types of foundations, and anticipated settlements are also usually included in the report.

c. Specifications - Specifications, even though generic in most instances, should be sought and accumulated since specifications may contain site specific conditions and construction procedures that may not appear in any other document.

d. Test Data - Data from field and laboratory tests performed during the project construction may provide valuable information about the materials used in the construction.

e. Miscellaneous Documents – Correspondence, change orders, contracts, field diaries, daily reports, requests for information (RFI's), applications for payment, and memoranda.

3.4.1.2 Applicable Codes

Review available drawings, specifications, and calculations for notes referencing design codes, material codes and other design guides. If no codes are referenced, determine the applicable code at the time of design based on the date of design documents.

3.4.1.3 Construction History

Documents retained by the Owner or Contractor during construction may provide insight into the difficulties encountered while constructing, and may provide tests and reports required by the design documents. Construction photographs, report logs, and engineering sketches detail the time line history of the project while providing a record of the "as-built" situations which may not be reflected in the original design drawings. The following items are generated during most projects, whether large or small:

- Soil Testing

- Bolt Tightening Tests

- Pile Driving Logs

- Concrete Compression Tests

- Steel Mill Test Reports

- Construction Field Sketches

- Photographs

During the lifetime of a structure, the Owner(s) may change or modify the structure many times. Some of the renovation projects will enlist proper engineering practices, but many projects are initiated and completed by local on-site staff and no documentation exists. The Investigator must review site-retained files for any modifications and compare them with the actual site investigation results. Tracking down the actual documentation is time consuming and may not be necessary if sufficient information is gathered during the field investigation.

3.4.1.4 Maintenance History

Maintenance records may provide another valuable resource. For example, reports of a persistent roof leak may explain corrosion or rot. An extended power outage may explain water damage where a sump pump was unable to operate during a period of heavy rain and flooding.

3.4.1.5 Shop Drawings

Shop drawings are usually requested by the Owner per the original design documents. Shop drawings may consist of structural steel fabrication, concrete reinforcement placement drawings, ductwork fabrication, piling details, and the like. Fabrication details most likely will differ from design drawing details because the original design drawings do not show complete detail information (especially for connections). To fully compare field conditions to the design drawing details, shop drawings should be sought. Again, this is often time consuming and may not be efficient, but shop drawings will more likely depict the actual field conditions during construction than the original design drawings.

3.4.2 Data Assembly

The final step in office work is to compile and document the prepared materials including sources of information, codes, standards, drawings, references, and assumptions.

3.5 REPORTS

The report is an important instrument of the Forensic engineer and report writing is an important part of their work. The art of report writing, like the art of all writing, concerns itself with ideas and style. However, a forensic technical report derives much of its strength from a factual and dispassionate presentation of its observations, analyses, results and conclusions. A typical outline of a written report on a failure investigation involves:

- Background

- Observations, including those made as part of a field investigation

- Information from others, including test results when performed by an outside laboratory

- Calculations and analyses

- Discussion

- Results and conclusions

The fact that four of these sections are written prior even to discussing the information overall is an accurate indication of the relative importance of these steps in laying the ground work upon which opinions will be based. A satisfactory report is based on exhaustive forethought and study. The preparatory steps described in previous sections of the set of guidelines represent at least four fifths of the job.

Where an investigation is part of a legal dispute, the forensic engineer's work product may fall within special rules of evidence. The forensic engineer should confer with the client and client's counsel as to the correct procedure for reporting interim results as well as final conclusions.

Failure investigation reports present the results of an investigation of the unknown. As such, it may seem that they are much more complicated and difficult than other types of engineering reports. This need not be so. Everything that has been said in the earlier sections of the set of guidelines regarding careful investigation, data gathering, data assembly, data organization, and analysis should support the forensic engineer in writing a clear and orderly failure investigation report. Each report should exemplify professionalism and reflect the highest technical standards.

3.5.1 Preliminary Report

When required, the preliminary engineering report, often called the PER, is often an information report, one reports that summarizes information and data gathered from the initial site visit but not one involving the discovery of new facts. As such, this is usually much easier to prepare than the final failure investigation report. The preliminary report preparation calls for no difficult explorations and no solutions to baffling problems. Only the gathering of known facts is demanded. The preliminary report may include a survey of facts, review progress to date, the steps used in the investigation, a description of the site, and any initial findings. Additionally, the report may outline plans or specific proposals for future research and investigation including any testing and consultants that may be required.

3.5.2 Final Report

The final report includes everything that needs to be said about the investigation including background information on the subject, descriptions of the various investigations, interpretation of results, findings, and conclusions. It should be very specific and describe exactly what was done and observed in the field, laboratory, and office. The language should be concise and leave no question in the reader's mind as to what was observed, analyzed, and concluded, and who did the work. The following sections provide one approach to organizing a final report.

3.5.2.1 Introduction

As its name implies, this section should introduce the reader to the subject of the report. In a letter report, the first paragraph may include preliminaries such as authorization, purpose, and scope. In a formal report, these may form part of the introduction.

3.5.2.2 Background

This section reviews the past history of the project and provides an overview of the failure including the following information:

- Brief history of the building/structure; including important places, names, and dates.

- Description of the failure and chronological account of the investigation.

- Previous researches and investigations which may have bearing on the present investigation.

- Extraordinary conditions that enter into the situation which must be understood before the specific questions of the investigation are approached.

- At a minimum, the background section should provide the history of the structure, the date of initiation and completion of construction, name and address of the owner, name and address of the general contractor, name and address of the architect and engineer of record, and a complete listing of construction documents. This section may end with a brief overview of the failure.

3.5.2.3 Observations

This section provides a description of the field investigation including the particular arrangement of the site, structure, objects, and other appurtenances and should explain clearly what was seen. This section should include only observations. The reason these observations are noted should be developed through analysis and discussion sections below.

3.5.2.4 Information from Others

This section documents the factual information not observed directly by the forensic engineer. This information may include notes on drawings, faxes, written and oral communication. In large studies, it may be appropriate to include in the body (or append separately) an exhaustive list of all documents reviewed. This section should generally include only those items that are relevant to the analyses, discussion, and conclusions of the report.

When presenting the results of laboratory work performed by others, the investigator should include only as much information as necessary to convey the results of those tests. If the full lab report is available or appended, this summary may be very concise.

3.5.2.5 Analyses

This section should review the types of analyses performed and report their results.

If the investigator conducted tests, these should be reported here in a manner consistent with general laboratory reporting. The description of a laboratory investigation demands a clear, impartial statement of materials acquired, methods employed, and results ascertained.

If the investigator performed computer analyses or hand calculations, these should be described in detail here. For computer analyses, include software name and release information.

For both tests and analyses, what is wanted is merely an exposition of facts. The order of such description may include the object of the investigation, the theory that underlies the analysis or test, description of the apparatus or method used, and the results.

3.5.2.6 Discussion

This section is the very kernel of the report. Here is where the forensic engineer crafts the story that ties everything together. The discussion incorporates the observations, the information from others, and especially the information gleaned from the various tests and analyses to inform the reader of the results of the investigation. For a structural failure, it seems obvious enough that at the time of the failure the loads exceeded the capacity of the structure. Here, however, the reader learns how the investigator established the likely loads, how the analyses and testing provide an understanding of the structure's capacity, and why those loads caused that structure to fail.

The findings or results of the investigation are presented in the order of their importance, the most important first. The inclusion of sketches is often helpful to support the discussion and to direct the reader to the important inferences drawn through the work done. The findings presented should then be used to develop the hypotheses described in this section. The language here should be concise and leave no question in the reader's mind. By the time the reader comes to the end of the discussion, the conclusions should be clear.

3.5.2.7 Conclusions and Recommendations

This section should list the conclusions in summary form together with any recommendations. Generally, this section should not include any new information, and should be a logical extension of the more extensive discussion above.

3.6 CONCLUSION

The failure investigation report is often written to serve as a basis for decision and action. The expert opinion of the Forensic engineer is required in the form of conclusions, and these conclusions are the distinguishing feature of the report. They must be reached only through the whole course of investigation, testing, and interpretation of results described in the body of the report. These require experience and sound engineering judgment and form the most important part of the report. It is important that the whole report should be so put together as to contribute to the logical, single end of giving weight to the conclusions. Photographs or images can and do have a strong impact on the report and its conclusion. Color photographs should be included, whenever possible, to aid in the description of the conditions found and conclusions reached.

The reputation of the Forensic engineer rests on the responsibility, practicability, and good judgment of the conclusions presented. Additionally, the Forensic engineer assumes some liability and should preserve an accurate record of his or her efforts by assembling and saving all records associated with the investigation.

3.7 REFERENCES

Ratay, R.T. 2000a. Editor-in-Chief, *Structural Condition Assessment Handbook for Serviceability, Code Compliance, Rehabilitation, Retrofitting, and Adaptive Reuse*, ASCE.

Ratay, R.T. 2000b. Editor-in-Chief, *Forensic Structural Engineering Handbook*, McGraw-Hill, New York.

CHAPTER 4 - ETHICS IN FORENSIC ENGINEERING

A truthful witness gives honest testimony and does not deceive.
- Proverbs 12:17, 14:5

4.1 INTRODUCTION

In practice, it has been found that much that goes on in courtrooms would be considered unethical by some engineers yet within-bounds by others. Most Forensic Engineers believe in the ideals of ethical practice, but have difficulty in recognizing when the line has been clearly crossed, either by themselves or by others. Lack of knowledge of accepted principles and canons of conduct, or pressures of the situation, can lead to a situation ethic, which may be appropriate or inappropriate. This chapter attempts to qualitatively and quantitatively define the line between ethical and unethical practices in Forensic Engineering by providing the precepts that circumscribe ethical forensic practices.

Litigation over technical or procedural causes of failure of engineered facilities has reached unprecedented levels in recent years. Forensic Engineers, enlisted in increasing numbers in these cases, are being retained to investigate these failures in order to provide courts with a rational explanation of the failure, in the form of investigation reports and expert testimony. Because of the adversarial nature of most failure investigations, Forensic Engineers may find themselves being pressured to champion their client's position rather than serving as an impartial purveyor of technical facts. At its extreme, yielding to this pressure may lead to inappropriate occurrences of advocacy, practicing outside one's competence, manipulation of facts, crafted testimony, inadequate or defective investigations, misrepresentation of standards of practice, and outright dishonesty.

Large numbers of engineers have transitioned to full-time work in litigation support. Many engineers advertise their availability for this work in attorney's trade association journals and on the Internet. While the number of engineers doing this is not a concern, nor is their choice of this clientele, their individual conduct is.

Ideally, the expert would both assist the trier of fact and help the client convey their understanding of the technical aspects of the case. Ratay (1997) points out that both NAFE and NSPE advise that the forensic engineer has a duty to their client and to honestly assist the client in efforts to establish where and to what extent liability lies. This conflict between assisting the trier of fact and the client's need for helpful testimony can create a serious ethical dilemma.

Accounts of questionable conduct have been reported in professional journals and publications and in complaints filed with state licensing boards (Carper 1998). These describe inappropriate, unprofessional or patently unethical conduct witnessed by others during their preparation for, or giving of, testimony. This behavior damages the profession and has led to increased interest in a better definition of the line between ethical and unethical conduct in Forensic Engineering. This chapter was prepared to assist engineers in avoiding advocacy or other unethical practices. No new or expanded guidelines of conduct are presented. Instead, the chapter compiles the guidelines contained in codes adopted by professional engineering associations and state registration board legislation.

4.1.1 Chapter Purpose

In its publication, "Ethics: Standards of Professional Conduct for Civil Engineers" (ASCE 2000), the society's history of development and interpretation of its code of ethics is documented. Regarding conduct, the document states, "ASCE is committed to the highest levels of ethical conduct." It also states, "To preserve the high ethical standards of the civil engineering profession, the society maintains and enforces a code of ethics." Thus, the yardstick for ethical conduct by Civil Engineers, as well as other professionals, is the code adopted by the respective association. This chapter on ethics is not codified, but presents guidelines of ethical conduct for Forensic Engineers derived from ASCE's code, as well as from numerous other similar documents. Though guidelines are given here, any evaluation of an engineer's conduct must be judged by the adopted code of that profession and not by these guidelines.

The premise of the entire chapter is summarized in ASCE's Code of Ethics, Canon 3.c (1996), which states,

> Engineers, when serving as expert witnesses, shall express an engineering opinion only when it is founded upon adequate knowledge of the facts, upon a background of technical competence, and upon honest conviction.

Codes of ethics for most other engineering professional associations contain identical or similar language.

The chapter begins by discussing the ethical issues encountered by Forensic Engineers, including descriptions of the application of ethics to some of the topics of Chapter 5, Legal Forum. The chapter then presents a compiled list of published precepts and principles of ethical conduct (see Appendix B) that apply specifically to Forensic Engineering. Though derived from published association and licensing standards, many engineers are surprised to find that the standards exist or that their practice falls outside the guidelines. A self-test of knowledge of the standards is provided in Appendix A. Next, the applicability of these precepts to each of the many facets of Forensic Engineering is explained. Finally, ethical aspects of the hypothetical example presented in Appendix A are described. A discussion of the damage done by unethical conduct, and appropriate methods of reporting unethical conduct, are included in the conclusion of the chapter.

Throughout the chapter, inappropriate practices are emphasized in order to help the reader distinguish between what is ethical and unethical practice. The term, "unethical" often connotes a tone of gross wrongdoing, which is not intended in this set of guidelines. Where used in this chapter, it should be considered to apply to the full range of wrongs from inappropriate, improper and even objectionable conduct to forbidden, illegal, or corrupt conduct.

Any discussion of ethics should note that errors are common to human behavior and are not ethical violations. Corley and Davis (2001) note that the cause of most failures is simple human error. Unintended ethical breaches result from carelessness or lack of knowledge of professional association or licensing industry standards and should be reported and corrected. Finally, it needs to be acknowledged that intended ethical breaches occur (Carper 1990). This chapter discusses published guidelines for ethical practice from adopted codes of conduct of engineering associations, and hopefully reduces the occurrence of unintentional breaches.

Due to its length, a list of 136 principles governing conduct of Forensic Engineers that was compiled for this set of guidelines is included in Appendix B. The sources used in compiling the principles are various codes of ethics and other documented guidelines described in the Appendix. The reader is encouraged to review the principles listed in Appendix B before reading the rest of this chapter, as many of the precepts are derived from the compiled principles. Studies by others (Lewis 1997) reveal that though these principles have existed for years, many Forensic Engineers are not familiar with the specific prohibitions and guidelines presented in the codes.

4.1.2 Ethical Forensic Engineering Practice Defined

Webster's dictionary defines "ethics" as, "the science of moral values and duties; the study of ideal human character, actions, and ends." A somewhat more applicable definition is given by the American Heritage Dictionary as "the principles, rules or standards governing the conduct of a person or the members of a profession." Ethical Forensic Engineering practice is defined in this set of guidelines as *"the conduct of forensic investigations and providing expert testimony based on sound, comprehensive, and unbiased investigation, and demonstrating exemplary, professional conduct and honesty in serving the trier of fact, the public, and clients, as a qualified expert."* Forensic Engineering practice outside this guideline is inappropriate.

4.1.3 Public Trust

All engineers, including Forensic Engineers, enjoy a public trust and are charged with responsibility in their practice for public safety, health and welfare. Failure in any way to recognize and fulfill this public trust constitutes an ethical breach. Thus, all engineers have the duty to the public, their clients, their profession, and themselves to embrace and apply responsible, ethical principles in their personal and professional conduct in engaging in any Forensic Engineering activities.

4.2 ETHICAL ISSUES IN FORENSIC ENGINEERING

Advocacy (defined below) and lack of competence, defined as alleging expertise in a particular subject area when the engineer knows he or she does not have sufficient expertise in that area, are probably the most prevalent ethical breaches in Forensic Engineering. Another leading form of misconduct occurs when the industry standards of care and practice are knowingly misrepresented to a judge or jury. Conflicts of interest, or semblance of conflict of interest, such as providing litigation support to an existing client (this is not considered unethical by the profession or in these guidelines, but can be made to appear improper by cross-examining attorneys), can also be causes for concern in Forensic Engineering. This section presents ethical aspects of these and a few other less prevalent issues of conduct in Forensic Engineering, and concludes with a discussion of liabilities of unethical Forensic Engineers.

4.2.1 Typical Issues

The Technical Council of Forensic Engineering has hosted a number of conferences on seminars on Forensic Engineering. These generally include presentations of papers or full sessions on ethical practice. A review of these deliberations and writings reveals that there are a number of recurring issues regarding ethics. The following list is representative of the issues identified by those working in this field:

- What constitutes ethical (and unethical) behavior by Forensic Engineers?

- Which code of ethics applies to Forensic Engineering?

- What's all the concern over advocacy (see discussion below)?

- How can an engineer serve both the client and the "trier of fact?"

- How can an engineer be an advocate of his or her opinion without being an advocate of one of the litigants?

- Is it unethical for an engineer who has full qualifications to "bend" the facts to support a particular point of view in the case?

- Is it ethical for an engineer to "emphasize" or "deemphasize" certain facts?

- What's wrong with contingency fees?

- What constitutes a reasonable degree of scientific (or engineering) certainly with respect to the cause of a failure?

- Should an expert witness volunteer relevant information even if the examining attorney does not ask for it?

- Can an engineer agree to give testimony in general on the subject matter without investigating the specific circumstances of a case?

- Is it unethical when an engineer won't concede indisputable facts?

- When is strategizing with the attorney appropriate? What constitutes "strategizing?'

- What constitutes "privileged" or "confidential" information?

- Is it unethical for an engineer to become passionately involved in his or her testimony?

- Is it appropriate to provide an attorney with cross-examination questions to be asked of adversaries' engineers?

- What activities or practices create a conflict of interest for Forensic Engineers?

- Do attorneys have a code of ethics governing their conduct with expert witnesses?

- Is it unethical for an engineer to testify if he or she has a perceived conflict of interest, even if it is not an actual conflict?

- What should Forensic Engineers do if they perceive unethical behavior by other engineering experts?

- Can an engineer testify on the quality of another engineer's services?

- Don't expert witnesses have immunity from liability when they testify?

- Is it always possible, and is it always appropriate, to abide by recommended practices of various professional organizations?

- Does the expert always have full control of objectivity?

Answers to most of these questions are directly or indirectly provided in this chapter, and not all represent unethical behavior. Some are completely appropriate.. The latter two are extremely important and were discussed by Ratay (1997) in regard to the ICED's *Recommended Practices for Design Professionals Engaged as Experts* (1988). Ratay describes some important caveats to strict adoption of the ICED's recommendations. Similar concerns with the thirteen ICED recommendations are described by Thompson and Ashcraft (2000).

4.2.2 Objectivity Defined

The primary task of a Forensic engineer is to investigate and ascertain the physical causes of failures and accidents. The legal and professional rules of forensic analysis require that the opinion of the professional engineer be the result of an objective investigation and unbiased analyses. Webster's definition of *objectivity* is adopted herein, given as "the state or quality of being detached, emphasizing the object, or thing dealt with, rather than the thoughts and feelings of the person." Black's Law Dictionary gives a similar definition. The technical investigation, and testimony regarding its findings, must be conducted objectively and independent of the views of the client. Failure of any Forensic engineer to uphold these high standards of objectivity is unethical.

4.2.3 Advocacy Defined

Webster's New World Dictionary defines an *advocate* as "a person who pleads or argues another's cause in a dispute." This definition, especially the phrase "pleads or argues" appears to be applicable to most objectionable Forensic Engineering advocacy and is the one adopted in this chapter. An engineer may discover that his or her client's position is completely upheld by the investigation, but must not assume or appear to assume the role of pleading or arguing the client's cause. This line is admittedly vague, as every engineer should maintain that his or her opinions are correct regarding the technical aspects of the case. The principles listed in Appendix B, plus material in the remainder of this chapter, should help define the line between advocating the client's cause and advocating one's own opinion.

Some Forensic Engineers see themselves as arbiters, having a responsibility to settle the issues. Black's Law Dictionary defines "arbiter" as, "a person chosen to decide a controversy; an arbitrator, referee." Under this definition, the Forensic engineer is not an arbiter, but merely voices an unbiased professional opinion for the trier of fact to consider along with other testimony.

Lewis (1997) identifies the issues in the objectivity versus advocacy debate as:

- should a line between ethical and unethical practice exist in the first place (i.e. what's wrong with fully advocating our client's position if an investigation supports its position),

- where should the line be drawn (isn't advocacy warranted at times on a case-by-case basis),

- how gray is the line (how can we tell when we cross the line), and

- has the problem grown to the extent that intervention is needed to protect the public and the engineering profession?

Advocacy is clearly prohibited by professional codes of conduct for Forensic Engineers (see Appendix B), but its existence and prevalence is well documented (Carper 1990).

4.2.4 Is There an Advocacy Problem?

Carper (1990) suggests that if a thorough evaluation of forensic practices could be conducted, it would probably disclose that advocacy is widespread. Some engineers are resolute in defending their right to use whatever means available to support their client's position in a dispute. Any engineer who takes this viewpoint meets the definition of an advocate and harms themselves, the profession, the public, and the court. The genesis of this viewpoint is probably centered in the fact that one side or the other of a conflict hired the witness, and both the client and most professionals understandably hope from the beginning that the client receives high value from the services. What must be remembered is that an objective investigation has value, regardless of whether it proves or disproves a client's position.

A common ethical dilemma occurs when engineers succumb to pressure by the competitive nature of the business, or by their clientele, to "go along to get along." Engineers' provide services, and their desire to excel in work in their field could result in succumbing to marketing, being recruited for, or accepting, advocacy work. Clients normally have a choice of engineering consultants, and may make a final selection on the basis of the one showing the most promise of meeting their craving for an advocate. Once selected, the engineer would encounter pressure to support their position, and may have a difficult time avoiding advocacy if the work was marketed with his or her up-front motives being other than providing objective investigations of the causes of a failure, leading to expressions of impartial opinions of the facts.

Zickel (1998) notes that a typical problem for all professionals performing as expert witnesses is the inclination of attorneys to attempt to convince the trier of fact that:

1. There is an answer to every technical problem in some book.

2. Virtually everything in the world of technology is a specialty that can be pursued as a course of study in some institution of higher learning.

3. Our education is so narrow that, if we have not majored in it, we are ignorant of the issue or subject.

4. Everything in technology is an absolute.

He relates a case where his client lost because he testified that distilled water readily absorbs soluble chemicals, which he had learned in high school. Though this was key to the proceedings, the jury was directed to disregard the testimony because he was not a chemist. His client's attorney failed to ask whether chemistry was included in engineering curricula.

An advocate was defined earlier as a person who pleads or argues another's cause in a dispute. Being objective implies that an engineer should not be interested in the outcome – defined as the verdict and award. Objectivity is served when opinions are based on evidence, sound technical principles, and judgment. Basing opinions predominantly on the client's position is advocacy. So is arguing or pleading the client's position, even if the opinions support the position. Regardless of whether an expert's investigation finds in favor of the client, the engineer has no business becoming, or appearing to become, one of the combatants.

Unfortunately there are so-called experts who will advocate the client's position regardless of engineering or scientific certainty. It is very important for Forensic Engineers to make it clear to their clients that they are not allowed by the profession's codes of ethics to be an advocate of the client's position. From the onset, the Forensic engineer's engagement must focus on performing an objective investigation to determine the causes of a failure or accident. The findings detailed in the investigation report may or may not coincide with the position or views of the client. This concept must be understood and accepted by the engineer, the client, and the client's counsel before the start of any investigation.

4.2.5 Solution to the Advocacy Problem

In light of the competitive and success-oriented nature of engineering services, how does an engineer take and perform an engagement without accepting, or giving an impression of accepting, the role of advocating the client's position? How can the engineer serve both the trier of fact and his or her client?

One promising method of assuring that the necessary and sufficient investigations to assure objectivity was suggested by ICED (1996), in which engineers who are approached by prospective clients would inform the litigant of the tests, investigations, or other research they would need to conduct in order to formulate their opinions, and to decline work when they are denied the ability to conduct all the investigations they believe are necessary.

Another approach to avoiding advocacy would be to carefully refrain from stating any form of preliminary opinion before conducting the investigation. It would also be beneficial if any announcement to utilize the engineer as a witness by the client or their counsel is withheld until after the investigation has been completed. This dispenses w ith t he p resupposition t hat t he e ngineer's i nvestigation w ill support the client's position, it prevents the client from prematurely announcing the expert's candidacy, it encourages involvement by qualified but reluctant engineers, it assures that the services are objective, and it allows the client access to the best opportunity to assess his or her liabilities and courses of action before including the expert on the witness list.

Though not likely to occur in our current justice system, the concept of a court-appointed expert has been discussed in various forums. This has a number of advantages but is complicated by several factors including distribution of costs, contacts and communications among the parties, and what criteria the court would use in selecting and receiving approval of an expert and of his or her opinions. Engineers use QA/QC methods to assure and control the quality of constructed facilities, and courts could implement OA/OC (objectivity assurance/objectivity control) plans to accomplish these objectives. As with QA/QC programs, both an assurance that the expert is being objective and control of that objectivity are required. A typical OA/OC plan could include the following elements:

bjectivity Assurance/Control Plan for Court-Appointed Expert Testimony

Plan Element	Element Description
imary Contacts	Each identified litigant will identify a single person to serve as the point of contact for all strategic discussions and correspondence with and by the expert.
n-strategic Contacts	Non-strategic, individual contact by the expert with any of the litigants or their counsel will be permitted but limited to data collection or explanation or clarification by the litigants of data, reports, or activities of the litigants. All contacts will be documented and distributed to the court and to the designated primary contacts for all parties.
eetings	Any needed meetings to discuss project strategies, scope changes or preliminary results will be announced to all litigants no less than 3

Plan Element	Element Description
	days before the meeting. These discussions will occur among a litigants in attendance. Notes regarding the topics discussed a conclusion reached will be distributed to the court and all litigants.
Correspondence	All litigants will receive all project-related correspondence develop by or received by the expert. During the investigation, any da reports and work products will be provided on request to any litiga Any time this occurs, identical information will be forwarded to t other litigants.
Site Work	Site tours of any facilities or field discussions of facilities will announced to the litigants no less than 5 days prior to the tour alo with invitations for all litigants to attend. The tour will proce regardless of the attendance.
Fiscal Management	The court will contract the fiscal aspects of the project with t expert, provided that they are willing to abide by the controls listed Items 1 through 5 above. The court will develop, outside of t experts' purview, agreements for reimbursement by the litigants.
Consensus	Following the data collection phase, the analysis tasks will reviewed and refined in collaboration with the litigants, and consensus on appropriate analyses adopted before proceedi Alterations of, or alternatives to, the contracted analyses will not permitted without unanimous approval of the litigants.
Litigant Comments	After completing the analysis, each litigant will be required to subm a written narrative description of how they believe the analysis affe the conclusions. These are not intended to be necessar scientifically based, which is the function of the experts' analys These "opinions" are requested primarily to disclose the factors t each litigant considers to be relevant and to assure that importa factors are objectively included in the final analysis. Becau different interpretations are expected from the litigants, a consens will not be required. The expert will include all of the interpretatio in the project report, and each will be compared with, and evaluat against, the expert's conclusions.
Scientific Approach	Wherever applicable, all conclusions shall be based on reasonal scientific certainty established from the application of approv scientific methods.
Predefined Conclusion	Prior to the analysis, the conclusion, such as the cause of failure, sh be predefined by stating what the conclusion will be based on t possible outcomes of the analysis.
Public Safety	If continuing adverse conditions are discovered, possible mitigati

an Element	Element Description
	measures will be identified by the expert. If a continued threat to public safety remains, mitigation of the risk will be proposed in the report.

Other methods of dealing with this dilemma, such as arbitration and mediation are described in the literature on Alternative Dispute Resolution (ADR).

4.2.6 Testimony Outside Area of Competence

Legal and professional criteria for establishing competency as experts were presented in Chapter 2, *Qualifications of Forensic Engineers*. Most engineers practice outside their particular specialty to some extent, but would probably not qualify as experts in all areas of their practice. Expert testimony should be limited to areas for which the engineer would qualify as an expert, as exemplified by the attributes listed earlier and repeated at the end of this chapter. An engineer who specializes in geotechnical engineering but had courses in structural design and hydraulics should probably not attempt to provide an expert opinion as to the adequacy or deficiency of the structural design or on hydraulic aspects of a bridge failure case. However, the engineer may suggest to the attorney that the structural design or unusual hydraulic conditions may be a contributing factor to the failure and should be looked into by other experts.

Black's Law Dictionary defines "incompetence" as, "lack of ability, legal qualification, or fitness to discharge the required duty - a relative term which may be employed as meaning disqualification, inability or incapacity and it can refer to lack of legal qualification or fitness to discharge the required duty and to show want of physical, intellectual or moral fitness." This definition embraces mental, moral, and physical incompetence, and is relevant here because a witness testifying outside his or her competence would be unfit to assist the court.

4.2.7 Inadequate Knowledge of Industry Standards of Care

Forensic Engineers are required to investigate a failure and testify as to whether applicable industry standards were "prudently" applied in the design and construction of the failed facility. This "care" in applying engineering standards requires that the engineer know the standards, especially if he or she participated in development of the standard, or through having applied the standards themselves. Engineers are allowed to evaluate and testify about others' applications of the standards. The Interprofessional Council on Environmental Design (ICED, 1996) recommends that engineers perform reasonable inquiry to identify the standards in effect at the time and place the procedures were implemented. It would also be fair for the expert to note whether standards have changed or to measure engineered designs by today's standards, but it would be inappropriate to judge a design by today's standards if the original standard was safe or if the standard has simply changed. It would also be wrong to condemn as unsafe or inadequate a design that met the standard of care that existed at the time the facility was designed unless the profession deemed the earlier standard to be unsafe.

Thompson and Ashcraft (2000) identify three problems arising from the requirement that experts testify only in regard to appropriate standards in effect at the time the failed facility was designed. They note that experts risk unintentionally raising the standard of care by relying more on their personal standards, they risk not knowing the standard of care at the time in question if they weren't in practice at the time, and they may be faced with situations were no standard existed. Five useful steps for defining the applicable standard of care are described by Thompson and Ashcroft.

Weingardt (2000) describes two generally accepted definitions of the standard of care. The first is that care which a reasonable person would exercise in a given situation with a certain set of facts, conditions, and circumstances. The second, more relevant definition calls for a comparison between the actions of a professional relative to the skill and learning ordinarily possessed by other reputable engineers in the community. Under both definitions, the court also examines how prudently engineers handled the application of the engineering standards.

Civil damage actions have resulted in identification of several forms of standards of care in engineering. Peck (1988) cataloged four applicable standards as:

- *Professional malpractice standard*, which holds that engineers are required to possess and apply the same skill and knowledge normally possessed by any members of that profession in good standing.

- The *duty to stay informed*, which dictates that engineers must stay abreast of common subjects of discussion among professionals working in the same field.

- A *strict liability standard*, which holds that the engineer's "product," whether services or some constructed facility, must be "fit" and without fault for its intended purpose. This is similar to an implied warranty for manufactured products. Courts have generally not applied this standard to engineers' services unless the engineer participated in the assembly or manufacture of the facility.

- A lower, *regulatory standard*, which states that an engineer who designs outside normal methods because it is encouraged by the owner, may have immunity from liability. The Environmental Protection Agency (EPA), for example, encourages consulting engineers to incorporate innovative methods in design by building immunity from liability, except in the case of gross negligence, into its contracts.

These encompass the range of standards of care that Forensic Engineers should know and apply in any engagements.

One problem is that standards that are eventually codified often lag new knowledge by many years. This issue has surfaced in pharmaceutical litigation, and courts are barring new and untested theories in lieu of accepted standards. The appropriate at-the-time practice codes, rather than current knowledge, should be applied in evaluating the performance of a constructed facility.

The expert should be familiar with the accepted practices that prevailed at the time the failed structure was designed and constructed. Unless required by law, an owner of a facility is usually not required to upgrade his or her facility to the most current accepted standard if the facility was designed in accordance to the accepted practices at the time of construction. To suggest otherwise would be inappropriate without proving that the facility is unsafe.

Courts determine negligence of designs based on whether the accepted practices at the time of design and construction of the structure was prudently applied. One of the requirements to ensure that the design was "prudent" is to be able to support it with adequate data, documents and analyses obtained and performed to assure that the design was in accordance with the accepted practices at the time of design and construction. It would be unethical to argue that a design was negligent or imprudently performed if it was comparable in quality to thousands of other similar designs of the time. This occurs frequently, but places the thousands of other designs, and the profession, at risk of losing credibility. The Forensic engineer may have exemplary hindsight, but would be acting unethically to presume that he or she would have foreseen events differently than the designer of the failed facility if faced with the same standards, and with similar understanding of the project risks.

Forensic Engineers must also embrace the above standards of care in their own practice. A Forensic engineer's investigation and presentation of opinions regarding a failure should be as "good" as services available from other experts in good standing in their field. Further, the work must be based on practical experience and state-of-art awareness, completely fit for its purpose of fact finding and assisting the court, and totally free of negligence.

A witness for the plaintiff may be tempted to exaggerate the standards of care to defeat the designer's defense, even though the design was within the limits of acceptable care. Alternatively, a defendant's witness may cross the line by inappropriately testifying that the failure was the result of circumstances not normally included in prudent designs. He or she may also attempt to re-define the standards in such a way as to press the bottom end of the standard just below the level of care used in the design.

Another common claim by witnesses is that "another" method or technique should have been used. This would be ethical only if the technique selected by the designer was within the acceptable standard methods in effect at the time the design was conducted. It would be unethical, however, for a witness to make this claim if the alternate method hadn't been tested as to whether it would have made a difference. The only ethical approach in these cases must begin with a clear, unbiased statement of the standard, followed by a description of whether the design falls inside or outside the standard approach.

4.2.8 Code Evaluations

Certain investigations are based on reviews of applicable codes and standards. Oftentimes a claim is made that the defendant violated a code or standard. This is an argument that is intended to prove definitively that the defendant was in error. The investigation either for the claimant or the defendant involves either a review of design calculations, construction plans or the site of the incident. A determination of what code or codes is applicable then has to be made. The investigator must judge what code is appropriate. Sometimes this is straightforward. The design of a structural member will be governed by local code. Today, most local codes in the US reference the American Concrete Institute (ACI) Code or the American Institute of Steel Construction (AISCP codes for details of design. What about older structures or structures built with systems no longer in use?

The investigator would be appropriately experienced with the code in order to evaluate its application. Is it appropriate for an engineer not licensed as a Structural Engineer in California to evaluate the seismic design of a California building? Some building codes are regional, such as BOCA. Use of the BOCA Building Code from one state to another may just involve selection of appropriate loading from design maps.

The matter may not involve a building code but a consensus standard such as American National Standards Institute (ANSI) standards or American Society for Testing and Materials (ASTM) standards. In these cases the expert has to show the thread between the design and the standard being cited.

The standard may not be referenced in a specific design code but may be selected as being appropriate by the Forensic engineer. There may not be any standard. The Forensic engineer uses his or her judgment to reference appropriate research, publications or possible standards. This is where responsible research by the investigator is most important. In many cases, there is a temptation to become an advocate by "reaching" for a possible standard in order to assist the client.

To summarize, in any code evaluation matter, the Forensic engineer should first have the appropriate experience with the code, should carefully evaluate the applicability of the code and should professionally determine applicable standards or guideline where no codes clearly govern the case at hand.

4.2.9 Conflicts of Interest

Conflicts of interest fall in two categories, actual conflicts and the appearance of conflict. An actual conflict occurs when there has been prior involvement, related to the current case, of the engineer with one or more of the other parties on a professional, personal or institutional basis. An *appearance* of conflict occurs when the engineer has had previous involvement with one of the parties on unrelated projects, or has rendered dissimilar positions in prior similar cases or in publications. In either case, the specific facts of the prior and current involvements determine whether the conflict is real. Apparent conflicts can be dealt with fairly easily, and are not a basis for discontinuing the engagement. Bias may be claimed due to prior involvements, and may exist, but does not make an engineer unethical. Failure to set the bias aside in favor of objective evaluation of the case at hand is unethical.

Actual conflicts and insoluble appearances of conflict of interest occur when there exists any influence, loyalty, interest or other concern capable of compromising the Forensic engineer's ability to provide an objective and unbiased professional opinion. ASCE (2000) defines a conflict of interest as arising in any situation in which a member uses his or her contacts or position in his or her employment to advance his or her private business, financial interests, or that of family and friends, whether or not at the expense of the client. They further state that members are expected to avoid any relationship, influence, or activity that might be perceived to or actually impair their ability to make objective and fair decisions.

Three degrees of conflict of interest for expert witnesses are implicitly defined in the ABA Code of Professional Responsibility (Johnson 1991). The first degree, or *actual conflict of interest*, are interests that are certain to adversely affect the opinion of the Forensic engineer. The second degree is *latent conflict of interest*. These are interests that may have a reasonable chance of affecting opinions of the engineer. The third degree of conflict of interest is termed *potential conflict of interest*. Potential conflicts of interest are involvements that can be foreseen to cause a conflict of interest.

One special form of conflict of interest is *bias*, which can be defined as an inability or unwillingness to consider alternative approaches or interpretations (Macrina, 1995). Bias can be eliminated from an investigation by utilizing different proven methods to investigate similar data, using a variety of accepted methods to analyze data and to investigate all possible scenarios and systematically prove or disprove their validity.

Before starting any forensic investigation, the investigation team should provide a full disclosure of all possible conflicts of interest to the client's legal counsel. The attorneys can then decide if there is a sufficiently high probability of conflict of interest that a different investigative team should be retained to perform the investigation.

Engineers provide services for remuneration. The last type of conflict arises where this interest is subverted, i.e. when an engineer cannot serve the public's interest and his or her own self-interest simultaneously. It is not valid to call a pecuniary interest a conflict, even though many cross-examining attorneys attempt to do so, because all experts need to be paid for their services. A pecuniary conflict exists when an engineer stands to gain more than the value of his or her services for their efforts.

4.2.10 Establishing Reasonable Degree of Scientific/Engineering Certainty

The phrase, "reasonable degree of scientific certainty" arises from courtroom rules of evidence, and as such is a legal term. Rather than elaborating on the legal definition, it is proposed in this set of guidelines that Forensic Engineers should rely on proven scientific and at-the-time engineering standards as the applicable means of assuring that their testimony meets this test.

One method of checking whether the "reasonable degree" test has been met is to consult publications, codes, standards and practice guidelines used by engineers working in the specific field of controversy. As an illustration, if three samples of a water supply source are needed by laboratories or regulators to establish whether contamination has occurred, a Forensic engineer should not testify on the basis of a single sample.

The legal standard, that of a reasonable degree of engineering certainty, suggests that all mutually exclusive modes of failures need not be investigated if evaluation of the most reasonable causes leads to a definitive conclusion. This may pass the legal test, but might fall short of the definition of a thorough investigation, or could hurt the case if the engineer's objectivity or bias in evaluating only one probable cause becomes an issue.

The measure of sufficiency of investigation is qualitative, not quantitative. This is not always the case, and an ethical investigator simply has to prove to himself the legitimacy of his results and be able to support his opinion by putting forth a sound technical foundation.

A proper forensic investigation should lead to the most probable cause of failure. To meet the reasonableness test, sufficient scientific basis must exist for each conclusion reached and for each failure mode rejected. Clearly, all possible avenues need not be fully exhausted in order for an investigation and opinions to reach reasonable conclusions. However, opinions based on insufficient investigation or erroneous data are inappropriate.

4.2.11 Failing to Resolve Disputed Facts

Ethical, opposing experts do not always arrive at the same conclusion, especially when engineering judgment is the cause of the difference. When experts' opinions differ, the examining attorneys and experts should clarify differences and the bases of the disagreements. Each expert should be interested in explaining the investigation performed and why it led to the conclusion made. This allows the jury to decide which explanation, and who's judgment, appears most convincing.

Actual disputed facts can in rare cases lead a judge to "give the technical case" to the jury. In so doing, the judge is acknowledging that the testimony by opposing sides was inconclusive because neither position was impeached; yet both cannot be true or meet the test of reasonable scientific certainty. The jury is instructed regarding the rules of engagement when this occurs. While this rarely takes place, its occurrence means that the Forensic Engineers failed to clarify and resolve technical issues, and the judgment of technical matters is left to the jury. In such cases, jury decisions can inappropriately place form over substance, and failure of an attorney or the experts to bring forth relevant facts may cause a jury to decide in favor of style of presentation by one side or the other.

4.2.12 Liability of Expert Witnesses

Forensic Engineers, like all other engineers, are performing a service within the practice of engineering. As such, the testimony itself can be subject to claims of negligence in preparing or providing erroneous testimony. Depending on the circumstances, engineers may not be immune from suits by those who relied on their expertise. Further, their liability might extend beyond their fee for services if, due to their errors or omissions, their client experienced damages such as a judgment involving substantial monetary penalties. Negligence claims against expert witnesses is a growing area of concern for the profession. Case law has imposed a legal liability on forensic witnesses if the expert deviates from accepted standards of care.

Some juries are willing to set aside the imputed immunity that expert witnesses have enjoyed heretofore, regardless of whether the judge acknowledges their expertise. If an engineer represents that he has a "greater skill or knowledge," he or she is required to exercise the skill and knowledge normally possessed by other experts in similar circumstances. As with any engineering service, expert testimony must be provided within the applicable standard of care. Other than what is contained in codes of conduct or licensing regulations, this standard of care may be difficult to establish. The principles tabulated in Appendix B are proposed as a list of precepts to assist in dealing with this question.

4.3 PRECEPTS OF ETHICAL FORENSIC ENGINEERING PRACTICE

Principles that guide or in many cases govern conduct of expert Forensic Engineering services fall in two classes:

- mandatory regulations found in contemporaneous codes of ethics or professional registration criteria, and

- those that do not necessarily govern engineers, but have been discussed by engineering societies as possible guidelines, or can be found in other references on ethics or civility.

Appendix B summarizes the ethical principles and precepts that apply to Forensic Engineering. The sources of the compiled standards range from state laws governing professional practices to codes of conduct contained in professional association criteria. This section starts by presenting a compiled list of principles found in the sources that apply to Forensic Engineering, then discusses the relevance to Forensic Engineering of state statutes and codes of conduct for several national professional engineering associations.

4.3.1 Compiled Principles of Ethical Conduct for Forensic Engineers

Ten categories of either mandatory or proposed principles regarding testimony and public statements by Forensic Engineers were identified by Lewis (1997) after reviewing just a few engineering society publications, professional registration documents, and other sources on civility and prudence. ICED (1988, 1996) list several categories of practice, some of which relate to construction disputes or reviewing work by other engineers, and are not relevant to Forensic practice. For purposes of this set of guidelines, the following 13 categories of Forensic Engineering ethical principles were selected:

1. *Forensic Engineers and Public Safety, Citizenship and Environmental Protection*

2. *Objectivity of Forensic Engineers*

3. *Competence of Forensic Engineers*

4. *Honesty of Forensic Engineers*

5.   *Thoroughness of Investigation by Forensic Engineers*

6.   *Relevance of Expert Engineers' Testimony*

7.   *Compensation and Business Practices of Forensic Engineers*

8.   *Conflicts of Interest in Forensic Engineering*

9.   *Confidentiality of Forensic Engineering Work*

10.   *Demeanor of Forensic Engineers*

11.   *Forensic Engineers' Conduct Toward Other Engineers*

12.   *Reporting Unethical Conduct of Other Forensic Engineers*

13.   *Professionalism of Forensic Engineers*

Canons pertaining to each of the above principles are listed in Appendix B. Sources of the principles and canons include the ASCE (2000) and NSPE (1985) *Codes of Ethics*, as well as ICED's *Recommended Practices for Design Professionals Engaged as Experts* (1988*)*, and ASFE/ICED's *"Recommended Practices for Design Professionals and Scientists Engaged as Experts for the Technical Review of Others' Work and Providing Testimony in Public Forums"* (1996). Wherever one of the canons in Appendix B is obtained from one of these sources, the reference is cited. All others were obtained from standard references on the subject. The reader is encouraged to review the list in Appendix B before continuing with the remainder of this chapter.

4.3.2 Knowledge of Professional Association Codes of Ethics

One of the quickest means of assessing one's own understanding of codes of conduct for Forensic Engineers is to review actual or hypothetical instances of testimony, and then try to assess the described behavior in light of published principles and precepts. Lewis (1997) developed a quiz that presented hypothetical statements by engineers, then listed principles that govern behavior of experts in providing such testimony. The paper discusses the need for and importance of engineers acquiring knowledge and complying with codes of ethics of our profession. Most engineers who take the quiz are surprised to find that professional and other related literature are not silent in regard to principles that govern or guide Forensic Engineering investigations and expert testimony. The illustrations in the quiz relate to drainage engineering, but the underlying issues and principles apply equally well in other fields.

4.3.3 State Statutes

All fifty states and four jurisdictional districts have developed statutes to regulate engineering practice. Many of the statutes also promulgate regulatory codes of ethics for the practice of engineering. The range of state regulations is too wide to summarize here. The state board of registration will provide these documents on request to any Forensic engineer considering practice in a particular state. For a national perspective, the reader is also referred to a recent, comprehensive survey of all state registration regulations (NSPE 1997). Data from this survey regarding *qualifications* of expert witnesses were summarized in Chapter 2. As shown in the survey, states have also developed a wide range of means of enforcement of the statutes and codes.

While all of the statutes differ to some degree, all without exception are concerned with ethical practice of engineering as it relates to public health, safety, and welfare. These requirements are the cornerstone of the engineering practice, and likewise, of Forensic Engineering.

4.3.4 Professional Society Codes of Ethics

All of the professional organizations concerned with engineering have developed codes of ethics. These codes contain many parts and are specifically concerned with the acts of individuals involved in the practice of engineering. Due to the large number of such organizations, a poll of their codes of conduct was not within the scope of this effort. However, two of the relevant association codes, ASCE and NSPE, were reviewed and are described here to illustrate the content. Each Forensic engineer has a duty to acquire and abide by the principles and canons of their respective professional and licensing associations. In the past, it was common to see a framed copy of the relevant code in almost any professional's office. These have disappeared in recent years, which is unfortunate, because posting the codes served to remind engineers of the importance of ethical practices.

The fundamental canons, rules of practice, and professional obligations listed in these codes embody, almost without exception, common principles of ethical conduct including:

- holding paramount the safety, health, and welfare of the public;

- practicing only in areas of the individual engineer's competence;

- issuing public statements in an objective and truthful manner;

- avoiding conflicts of interest; and

- avoiding deception in the solicitation of work.

Most of these codes provide examples of proper conduct.

The legal implications and mandates of these ethical codes of professional organizations has created a minimum standard of practice, including ethical principles, to which all members of the associations are held.

4.3.4.1 ASCE Code of Ethics

For background on ASCE's codes of ethics, the reader is referred to a comprehensive review by the society's General Counsel (Smith 1999). The American

Society of Civil Engineers adopted its original code of ethics in 1914 and has updated it in 1927, 1961, 1964, 1975, 1977, 1980 and 1996 (ASCE 2000, Smith 1999). The only specific mention in the current code of conduct while serving as expert witnesses is in Canon 3.c. Other principles in the Code that peripherally apply to Forensic Engineering have been tabulated in Appendix B.

Canon 3.c of the ASCE code relates to three ethics categories, thoroughness of investigation, competence, and honesty. It states,

> *Engineers, when serving as expert witnesses, shall express an engineering opinion only when it is founded upon adequate knowledge of the facts, upon a background of technical competence, and upon honest conviction.*

The current ASCE code has been reviewed for relevance to today's practice (Groden 1991, Smith 1999) and by the TCFE Ethics Committee (TCFE 1992). TCFE examined the Code relative to Forensic Engineering and assessed its applicability to the future. The committee determined that the Code *was not only adequate but also exemplary in its definition and objective.* They concluded that Canon 3.c was the principal operative for Forensic Engineers, and that unethical practice is usually traced to the lack of adequate knowledge of the facts or standards, lack of technical competence, or lack of honest conviction (TCFE 1992).

It is interesting to note that TCFE's Ethics Committee also concluded, in the context of competence, that testimony by university professors and research scientists who do not have a background in forensic practice should be restricted to their narrow area of expertise and should exclude statements regarding the cause of failure. This is not a statement of ASCE policy, but communicates the concern by many design practitioners that a prerequisite for testimony regarding failure of designed facilities is experience in design. Any engineer who is considering an engagement as an expert, whether associated with a university or research center, should first review the qualifications described in Chapter 2, and then consider the ramifications of testifying in regard to design if their background has not included designs or exposure to design standards.

4.3.4.2 NSPE Code of Ethics

The National Society of Professional Engineers, an organization shared by engineers from all facets, adopted a revised code of ethics for engineers in 1985. In addition to addressing rules of practice and professional obligations of members, it incorporated provisions to reflect the April 1978, U.S. Supreme Court decision that the Sherman Act does not require competitive bidding for engineering services on federal projects. It contains expanded definitions of professional relations, standards of integrity, public interest, deception, conflicts of interest, unfair competition, injury to other engineers, and continuing education. The code is available by contacting NSPE or any of its local chapters.

Principals that apply to Forensic Engineering from the NSPE code were incorporated in the 13 categories listed in Appendix B. In general, it was found that the code adds to rather than duplicating the principles applicable to Forensic Engineering in the ASCE code.

4.3.4.3 ABA and ASCE Positions on Ethical Aspects of Civil Justice Reform in America

The 1990 President's Council on Competitiveness developed a series of 50 recommendations for civil justice reform in America. The list, along with responses of the American Bar Association's (ABA) Section of Litigation's Counsel, was published in *Litigation News* in December 1991. For a period of years, the Technical Council on Forensic Engineering (TCFE) existed within the auspices of ASCE and served as a forum for development of ASCE's positions on Forensic Engineering issues. The ethics committee of TCFE (1992) determined that six of the Council's 50 recommendations pertain to expert testimony. After reviewing the ABA responses, TCFE e ndorsed a ll s ix r ecommendations a nd s econded ABA's responses. The six recommendations and ABA's responses are shown in Table 4-1.

The "market incentives regime" mentioned in recommendation 22 refers to an earlier recommendation (No. 7, not listed) related to placing quantitative limitations on discovery and discovery costs. Neither the ABA nor TCFE supported the limits proposed in recommendation No. 7.

These recommendations are included here because they represent positions by ABA and ASCE regarding several aspects of Forensic Engineering. Though not adopted as policy, they should be viewed as guidelines for ethical practice in the six topics. In particular, the ban on contingency fees endorsed in item No. 20 is applicable to any discussion of ethics.

In addition to codes and proceedings of professional registration boards, numerous other engineering publications incorporate guidelines for conduct by expert witnesses. Though not necessarily endorsed by the society, papers and articles on the s ubject receive peer review and are accepted for publication to inform the membership. Many of these are controversial, but indicate that the subject is important to the society. For example, Babitsky (2001) addressed the issue of volunteering information during courtroom testimony and suggested that "contrary to what may be appropriate in other circumstances, experts should not seek to help the counselor by volunteering information to clarify the issues." The paper elaborates on the writer's reasons for this statement, but it leaves many questions regarding the expert's principal role of assisting the trier of fact.

Table 4-1 American Bar Association Recommendations Pertaining to Expert Testimony

Recommendation (Endorsed by TCFE 9/12/92)	ABA Responses
19. Require expert testimony to be based on "widely accepted" theories. A party	SUPPORT reform in the area of expert testimony. SUPPORT use of a reliability'

would have to prove that its expert's opinion is based on an established theory that is supported by a significant portion of experts in the relevant field.	standard rather than 'widely acceptable' standard. NOTE that the civil justice reform recommendation fails to address the parallel issue of expert evidence in state courts.
20. Ban contingency fees (compensation in return for a 'successful outcome') for expert witnesses.	SUPPORT this recommendation.
21. Permit more comprehensive inquires of proposed 'expert' witnesses through interrogatories and disclosure of additional core data, including the expert's publications and compensation arrangement.	SUPPORT additional comprehensive inquiries of proposed 'expert' witness.
22. Additional expert discovery such as depositions for expert discovery should be permitted, subject to the market incentive regime outlined above.	SUPPORT expert deposition discovery without need for court order. However, OPPOSE so-called 'market incentive' regime except to the extent that the current system provides that requesting party must pay for experts' deposition time.
23. Require courts to determine that proposed expert witnesses are legitimate experts in their field before they are permitted to testify.	SUPPORT courts' determination of whether proposed experts are qualified to testify. Requirement that experts be 'legitimate experts' in their field may be too narrow and limit access to the courts.
24. Resist attempts to take the review of expert testimony away from juries or force the use of court-appointed experts.	SUPPORT the resistance of attempts to take away the review of expert testimony from jurors or force the use of court appointed experts.

4.4 ETHICAL PRACTICE GUIDELINES FOR FORENSIC ENGINEERS

Forensic Engineers are not just expert witnesses. They are businesspersons, technicians, investigators, communicators, professionals, and engineers. Forensic Engineers find themselves involved in all phases of professional practice from solicitation of work to final close out and retention of records of projects. Ethical practices are mandatory in all phases, not just the portion involving appearances in court as an expert witness. In the order of occurrence of work phases, this section describes ethical considerations that should be given to each stage of Forensic Engineering practice. The business aspects of each phase are discussed in chapter 6.

4.4.1 Solicitation of Work

State registration boards generally promulgate rules regarding allowable forms of solicitation of work by professional engineers. Codes of ethics for professional associations also address this topic. Generally, advertisements placed in phone books, trade journals or professional society publications are appropriate if professionally and truthfully presented. Some engineering firms advertise in newspapers, radio and television ads. The medium of the solicitation is not as important as the content when ethical aspects are being measured.

Attorney's looking for "expert" witness engineers can find a plethora of ads from engineers who are eager to testify for their clients. *Trial*, the magazine of the Association of Trial Lawyers of America, often contains three or more pages of "Engineering" classified ads each month. These unfortunately often include questionable clauses such as the engineers' courtroom experience with statements like, "over 50 expert testimonies," or inviting readers to call for a "free assessment of their case." Only about half the ads by engineers indicate whether they are licensed. The National Academy of Forensic Engineers has taken issue with engineers who actively advertise for opportunities to provide courtroom testimony, particularly if they make such appearances their primary source of income.

In addition to advertisements in trade magazines, Internet and e-mail ads for litigation support services are appearing in greater numbers. Entry of the two-word phrase, "Forensic engineer" in most search engines produces tens of thousands of "hits." Advertising companies also take liberties that approach or cross the line into unethical areas.

4.4.2 Initial Client Contact

The initial client contact will most frequently occur by telephone. Regardless, the first interaction will establish the pattern for an ethical versus improper relationship over the duration of the engagement. Borderline practices, often accidentally employed during the initial contact, may mislead the client and invite continuing problems in other phases of the work.

The Forensic engineer should have ethical objectives and principles in mind, as well as business and technical considerations, during any initial contact. His or her attitude and tone should portend objectivity, competence, thoroughness, professionalism, financial integrity, and honesty.

If the work is within the Forensic engineer's expertise, the engineer will naturally want to provide the services. However, this goal should not control the direction or outcome of the contact. Nor should the discussion focus on ways that the expert could support the client's position because it would not be possible without an investigation for the engineer to make this conclusion. Early in the contact, the engineer should strive to establish whether the topic is in his field and that the client is willing to authorize a thorough, objective investigation. The client should also be willing to concede that the decision to provide testimony in regard to the findings will occur after the investigation reaches its conclusion. Compensation for the services should be discussed, especially the need to affirm that the client is willing to pay for the investigation even though it may conclude that the client may not prevail in court. Any potential conflicts of interest should be disclosed and a discussion of their impact should occur prior to setting the final terms of the contract.

The Forensic engineer should also obtain background information such as:

- What happened to occasion this call?

- How did the caller obtain the engineer's name?

- What are the circumstances surrounding the failure?

- What is the relationship of the caller to the failure?

- What are all the perceived technical issues, and are they totally within the Forensic engineer's expertise?

- Does the caller plan to engage other Forensic Engineers in the same field? In other fields?

- What level of effort is expected, and what budgetary or schedule limitations is the caller faced with?

- Has the caller considered terms of engagement?

- How and when will the engagement get started?

- When will preliminary results of the investigation be needed?

- If an attorney makes the call, will his client handle the contract?

- What would the caller like to have to further evaluate the engineer's credentials?

The initial contact should also ascertain the next step. Should the engineer send a resume, return a confirming letter, be available for an interview, volunteer to research some of the case circumstances before further negotiations, propose terms of a contract, or schedule a meeting to obtain more details of the case and discuss qualifications, scope, schedule and any conflicts? As this is usually the initial communication, the engineer's ethical patterns, whether good or bad, are demonstrated here and normally carry through the rest of the engagement.

4.4.3 Contracting for Services

Care in development of the agreement or contract for the investigation is needed to avoid several possible unethical traps. Just as limited-services contracts in design work can result in overlooking important factors that often lead to failure (TCFE 1992), composing an agreement for an investigation that will attempt to prove the client's position can result in overlooking other causes of the failure. Other pitfalls in contracting include:

- Associating compensation provisions with the outcome of the investigation

- Confusion regarding compensation amounts for appearances (depositions, testimony) versus investigation costs.

- Including a scope of work that has appearances of advocacy.

- Underestimating the level of effort needed to conduct an appropriate, thorough investigation.

- Including unrealistic schedules for completing a proper investigation.

- Accepting a specific, limited scope as the only likely effort that will be contracted.

Numerous textbooks on expert witness work, such as Feder (1991), contain sample engagement letters, sample scopes of services, and sample contract forms.

4.4.4 Scope of Services

Every contract needs a scope of the basic services, plus provision for additional services should the client request them. Care should be taken in developing the scope of services to avoid language that implies that the work is not thorough or that it appears to focus only on the client's point of view. Time constraints, fee constraints, and availability of the engineer may limit the final scope, but sufficient resources in all these categories should be incorporated to assure an adequate analysis of the failure and its causes. The scope of work may be discoverable, and should be free of language that suggests that the work be crafted to produce a particular conclusion. Feder (1991) recommends that the following elements be included in a typical scope of services for a forensic analysis of a failed facility:

- Planning meeting

- Analysis of claim

- Discovery, on-site and off-site

- Document collection, review and indexing

- Project analysis

- Analysis of facility construction schedule (if the failure involved a constructed facility)

- Review contractor's bid and claim documents

- Litigation support

- Meetings and Reports

Though not included in Feder's list, the engineer should also analyze design drawings, specifications, shop drawings, inspection reports, and records of job meetings.

Since only a sufficient investigation will lead to opinions meeting the reasonable-degree-of-certainty test, to agree to perform less is highly unethical. An opinion that is based on insufficient information is easily undermined and does not serve the client's interest or court's needs.

4.4.5 Compensation

The compensation an expert receives cannot be linked to the outcome of the investigation or the outcome of the trial. Attorneys often accept a case on a contingency fee basis, but this is inappropriate for engineers. Without exception, contingency fees for engineering investigations are prohibited by the codes of practice of professional associations and state licensing boards.

Considerable debate exists as to whether contracts should be implemented that call for compensation by a law firm, especially when working on the plaintiff's case. Though the client's attorney might be the primary point of contact during the investigation, a perception of conflict of interest may occur if the engineer's compensation is paid by the law firm rather than the client. The engineer must assure that his or her compensation is not dependent on a favorable judgment on behalf of the client. It may also be considered unethical if a contract is implemented that calls for attorney approval of the engineer's invoices prior to payment by the client.

Practically every engineer is asked, on cross-examination, "are you being paid for your testimony?" Experts must be cautious in answering to avoid an appearance of bias or a conflict of interest. ASFE (1987) recommends the answer, "I am being paid for my time, not for my testimony." Most accomplished experts find that an honest, unrehearsed response is adequate.

4.4.6 Confidentiality

If information of a confidential nature is provided to an expert by his or her client, the expert should keep records of the consultation and honor the confidentiality. An expert should also notify the retaining lawyer if third parties attempt to discuss a case with the expert.

Communications between the engineer and the client's attorney are normally considered to be confidential and privileged, and are usually not subject to being produced during discovery. This rule applies to correspondence or exchanges of written materials that are a part of the overall development of a position, and possibly of a strategy for handling the case, by the client and their attorney. Little concern exists here regarding ethical practices, with the exception that the engineer retains all materials, whether they've been marked or identified as confidential materials. Wise choices in what materials are retained in paper or computer files usually obviates any difficult decisions regarding what materials are to be produced in response to an interrogatory or other discovery mechanism. With the exception of privileged materials, the ethical engineer has a duty to produce all the contents of the project files if asked for these items.

4.4.7 Conducting the Investigation

Ethical considerations should carry over from initiating the contract to implementing the investigation. Unethical practices in this phase could include:

- Removal of evidence from the scene.

- Altering evidence at the scene.

- Deleting or redacting non-supportive notes or information from documents, publications or exhibits.

- Deleting portions of data from records.

- Altering or discarding photographs or videotapes.

- Withholding relevant material during discovery.

- Being vague about work being conducted, or in stating opinions.

Failure to conduct a thorough investigation of all reasonably potential causes and effects can result in conceivable erroneous conclusions, or as a minimum, loss of credibility when other reasonable causes are discussed by the opposing attorney while the witness is on the stand. Without thorough investigation, an improper linkage of cause and effect can be presumed, or other possible causes may have been neglected.

A good example of an inadequate cause-effect analysis occurred in the 1998 lawsuit by Texas cattlemen against Oprah Winfrey. Their claim was that her televised show on "mad cow" disease caused cattle prices to drop after the show from $61.90 to $55 per hundredweight. The witness for the cattlemen, a professor from a leading university, stated his opinion that the show's impact depressed cattle prices for at least 11 weeks. Under cross-examination it was learned, and had to be conceded by the witness, that cattle prices usually drop in the same season that followed the show. Winfrey's defense lawyers claimed that the prices could have easily dropped because of season, drought, poor exports, oversupply, or other negative media attention besides their client's show. Of ethical relevance here is whether the cattlemen's expert witness investigated the other possible causes in reaching the conclusion.

Even though the client/attorney objectives of proving their viewpoint and winning the case are important, it is paramount for the engineer to consider as many reasons for the failure as possible. The engineer should not limit the investigation to causes that are beneficial to his client. Most failures are usually due to more than one reason, and often one failure mechanism could trigger another.

The scope of the forensic expert's planned investigation should call for evaluation of all reasonable causes. An attorney or client may desire a scope that is limited to investigating a single cause, which may be appropriate but should be debated in the presence of the attorney early in the engagement. If there is a scope conflict between the attorney, client or the expert, it should either be resolved early or the expert should not take the assignment.

4.4.8 Report Preparation

Oral Reports - Oral reports are usually made during the initial portions of an investigation and are probably the most appropriate means for reporting one's preliminary observations.   These should not include final conclusions unless the investigation is final and the opinions are clearly and conclusively founded.   Even with substantiating data and analyses, it is best to limit statements of conclusions or opinions to written reports so that the most appropriate means of wording the opinions can be set by the engineer, leaving only the means of disclosing the opinions to the client's legal counsel.   Feder (1991, p.223) provides an excellent outline for preparing and giving a preliminary oral report.

Preliminary Written Reports - The preliminary engineering report should be prepared only after collecting the needed data, reviewing it and completing the investigation of the facts of the case. Care should be taken in writing the report and discussing it with the client or attorney to state the primary findings without elaborating on implications of the findings or strategizing the presentation of the case.

Final Reports - The final report should identify the steps and procedures followed in completing the investigation, including names and dates of people contacted.   It should contain the relevant supporting data and summaries of calculations.   All possible causes of failure that were investigated, and reasons that others were not investigated, should be disclosed.   Once the investigation is described, it should clearly state the forensic expert's opinion as to the causes of the failure and imprudent applications of standards, but should not state opinions regarding guilt or negligence, as this is the role of the court.   Applicable standards of care should be attached and referenced in the narrative.

4.4.9 Attorney's Reviews

It is acceptable to allow the client's attorney and the client to review drafts of preliminary and final reports. The reports should be forwarded, under client/attorney privilege, to the attorney rather than the client. The attorney then forwards the report to the client under privileged correspondence. It is acceptable to forward the report as a draft, and to make changes in the presentation of the facts and findings after receiving comments from the attorney and client. It is inappropriate to modify the opinions at the request of either.

4.4.10 Strategizing with the Legal Team

Providing explanations of technical matters, aiding in presentation strategy and general consultation regarding the technical (not legal) case are all services ethically performed by Forensic Engineers. What is unacceptable is participating in the "win" strategy, or altering or crafting opinions to fit the strategy. Presenting highly credible witnesses is the principle legal strategy of most ethical attorneys. Because credibility is the hallmark of the ethical expert witness, the objectives of both can be completely compatible. There is nothing unethical about Forensic Engineers involving themselves with attorneys in developing and presenting a credible technical case.

4.4.11 Disclosure

Changing rules regarding disclosure are trending toward assisting courts in obtaining information regarding qualifications of witnesses and their opinions without going through the lengthy disclosure processes used in the past. Forensic Engineers should be aware of the purpose and techniques of disclosure, and avoid becoming part of the strategizing and positioning that are prevalent during this process. Rather than engaging in the strategies, the engineer should be prepared to respond to interrogatories or depositions in an ethical, professional manner.

Potential ethical breaches in disclosure during discovery include:

- Excluding pertinent background or previous experience information because no one asks for it.

- Failure to explain who did the actual investigative work.

- If another engineer did the work, assuming that the testimony would be accepted based on the reputation of the witness alone.

- Willfully ignoring factual data.

- Attempting to recant prior testimony or published positions because they do not fit the client's theory.

- Deleting references to professional-society involvements from a resume.

- Crafting the resume to mislead the reader.

- Deleting certain previous engagements or publications from a resume.

An example of the last item occurred in the Oklahoma City bombing trial of Terry Nichols. An expert witness for the prosecution was subsequently grilled by the defendant's attorney when it was discovered that the witness' resume had excluded mention that he had also served as a ballistics expert for the government in their investigation of the Ruby Ridge standoff. When asked why he left it off his resume, the prosecution objected and the judge sustained the objection. The jury didn't hear the witness' reason but they certainly heard the question.

Experts should insist on including all pertinent information in disclosure, even when it is or may be adverse. One engineer, when asked if full disclosure was appropriate answered, "no, that's why we have two sides." It is not the duty of the other side to obtain a thorough resume from the opponent's witnesses, and it is unethical to fail to cooperate in both sides' need to receive all relevant background information.

4.4.12 Public or Professional Statements

Any public statements regarding work or services being performed should be avoided, and should never be given without approval of the content by the client or their counsel. Publication of results of the investigation should not occur until after the case is resolved and all appeals are completed, and then only with client approval.

Other parties may contact an engineer during an engagement, but questions should be referred to the attorney handling the case.

4.4.13 Testimony at Depositions

The form of questions asked during depositions is often different than during trial. Many lawyers save narrow, carefully phrased questions for the trial, while depositions are a time for open-ended, general, "why" types of questions that have unlimited explanations. The opportunity to add full explanations to answers during depositions can result in temptations to exaggerate, include irrelevant material, or intentionally preclude deeper inquiry into important aspects that have less-supportive, factual foundations. Yielding to these, or other similar temptations that have surreptitious goals, is unethical. On the other hand, witnesses may also attempt to lay expanded explanations in the deposition record, anticipating that their client's attorney may want to use the deposition on redirect during the trial to admit lengthy explanations that would not normally be allowed on cross examination.

4.4.14 Testimony at Trial

The greatest opportunity for ethical breaches occurs during testimony at trial or in hearings. These range from outright untruths to less-damaging breaches that can injure the proceedings and the profession. Though some of the following may be appropriate depending on circumstances, the most common less-damaging actions include:

- presentation of extraneous information,

- speculation,

- unfair generalizations,

- incomplete treatment of data,

- irrelevant statements,

- avoidance of pursuit,

- inconsequential errors, and

- erroneous testimony.

Extraneous information should be excluded and data relevant to the argument should be stressed. Speculation is the inclusion of unnecessary and insupportable statements to explain data. Unfair generalization is the use of statements that do not specifically apply and are an attempt to unfairly include the failure in a category to which there may be exceptions. Use of terms such as "generally," "usually," or "in most cases" imply exception and are discouraged if they do not serve the court's interest in specific applications to the case at hand.

Speculation occurs when conclusions are drawn beyond reasonable degrees of scientific certainty. Every expert is allowed to formulate and state opinions, but should not cross the line of reasonable degree of certainty. Many breaches occur when engineers provide opinions outside the foundation laid, often because they feel they can and will not be discovered or reprimanded. This practice is not serious if the judge or opposing attorney is aware of the line between exact and inexact science, because they can object to the testimony and have it stricken. However, the expert has the duty to keep from crossing this line, rather than expecting the court to refrain him or her, because he or she has greater knowledge of the technical limits than court officials do.

An example of this occurred in the Oklahoma City bombing trial of Terry Nichols. A government expert testified that the drill bit found in Mr. Nichols' house matched the markings on the padlock at a quarry where blasting caps and sticks of explosives were stolen from a storage shed. When the expert testified that he believed Nichols' bit specifically matched the padlock marks, the defense objected on the basis that such matching is inexact. The judge, who had researched the subject months earlier, sustained the objection and ordered the jury to disregard the testimony. The witness was allowed to show jurors the similarities between the bit and the drilled hole, but not to tell them he believed Nichols' bit made the hole. If the witness knew that exact matching was not scientifically possible, he crossed ethical lines when he testified that the bit made the hole.

Incomplete treatment of data occurs when data presentations invite questions for which no answer is presented or prepared. Irrelevant statements are any arguments that have no potential for establishing facts but are often offered to influence conclusions. Avoidance of pursuit occurs when a witness gives long-winded answers that make the answer to the question almost indiscernible or fail to answer the question by rephrasing the question.

Inconsequential errors such as incorrect references to figures, incorrect bibliographic references, typographical errors, or transposition errors are not relevant to the arguments but can cast doubt on the thoroughness and care of the engineer. Erroneous testimony does the most harm if unchallenged because the lay judge or jury are inclined to accept it as factual. Errors can occur in understanding of concepts, in calculations, and in data presentation.

Besides those mentioned, other borderline or unethical courtroom practices of engineers may include:

- Volunteering answers or information beyond the scope of the question.

- Exaggerations.

- Pontificating.

- Criticizing or demeaning another engineer's work.

- Punishing the cross-examiner by giving lengthy dissertation answers.

- Fencing with the attorney or judge.

- Giving opinions not asked for.

- Inspecting exhibits prepared for the proceeding by others without their permission.

- Discussing testimony during recesses with persons other than the client or client's counsel.

- Openly displaying approval or disapproval during other witnesses' testimony.

Use of showmanship, puffery, jingles or sensational language is unprofessional and inappropriate. An illustration of this appeared recently in a published account by an engineer (undisclosed reference) who felt he had done a good job. It describes his first encounter with expert testimony. The discussion during cross-examination about whether parts from a camera could be used in igniting a bomb went like this:

Q: "You know nothing about electronics, right?"

A: "I'm not a bozo with big shoes or a red nose."

Q: "You have to be a genius to trigger a bomb, right?"

A: "Fraid not. Any moron will do."

Q: "Flash tubes mean cool flash, so there's nothing there to trigger a bomb, right?"

A: "Sorry, lots of energy there to ignite things. Ruins your whole day if you short out the flash capacitor with your tongue."

Though this type of testimony by engineers sometimes occurs, and often with pride as was the case here, it cannot be considered to fall within the NSPE (1985 and 2001) code of conduct (Canon III.3.a) or the precepts of ethical conduct listed in Appendix B.

4.4.15 Dealing with Adversarial or Confrontational Situations

Webster defines an adversary as, "One turned against another or others with a design to oppose or resist; an antagonist; enemy; foe." The determination that those of differing opinions or judgments are adversaries is unethical and should not be part of an expert's stance. Although "side-taking" serves well for opposing legal counsel, an adversarial relationship between engineers or between an engineer and an opposing attorney can encumber the activities of those performing Forensic Engineering services.   An adversarial relationship can compromise objectivity, compromise professionalism, or damage the witness' credibility.

When another perpetrates an adversarial relationship toward an expert witness, the Forensic engineer is well advised to avoid direct confrontation over the differing opinions or positions. This rule of *avoiding* rather than engaging in confrontation also applies to other relationships, but is of greatest importance during adversarial exchanges.

Methods that the Forensic engineer can use to avoid direct confrontation include:

- Establishing common ground (it may be as simple as agreeing that both parties are doing the best work that they can),

- Remaining pleasant and calm,

- Making it clear that professionalism and integrity are core to the analyses conducted and the opinions given, or

- Re-directing the issue under discussion to a form that is less confrontational.

When confrontation becomes unavoidable, the following results may occur:

- Legal counsel becomes a more active participant in the process,

- Information becomes more difficult to elucidate,

- Demands and deadlines become harsher and, at times, unrealistic, or

- Lines of communication between parties become strained and possibly severed.

In addition, the Forensic engineer confronted with hostility should:

- Hold his or her ground in a calm, steadfast, and resolute manner,

- Control his or her emotions and resist responding in anger (this type of response will only elevate the confrontation), or

- Take his or her time in listening and responding (time and periods of silence are powerful tools in disarming confrontation).

The Forensic engineer should not engage in written or spoken sparring episodes. Attorneys often duel, but this is not appropriate for the Forensic engineer.

4.4.16 Assisting with Remediation

Before, during or after a trial, a Forensic engineer may be asked by the court or client to assist with remediation of the failed facility.   This can occur during the investigation phase, but usually transpires after a judgment is rendered and the decree requires immediate remediation of and unsafe condition. Because remediation normally follows a decision, opportunities for inappropriate behavior decline dramatically, but deserve the same care in avoiding ethical breaches. Three instances are usually encountered as follows:

• Where remediation has already occurred:

In the cases where remediation has already occurred, the Forensic engineer may be asked to ascertain whether the remedial actions taken were sufficient. Only when the Forensic engineer has formed an opinion as to the causes of the failure and independently analyzed whether the remediation meets design standards can an opinion as to the sufficiency of the remediation be rendered.

• Where remediation has not yet occurred:

If remediation has not been applied, the causes of the failure must first be ascertained. Only when the causes are known can remediation, specific to the failure, be designed. The expert may be called upon by the court to cite appropriate standards for design of remediation, or may be asked to design the remediation facilities. An ethical dilemma that arises in this instance is when a court orders or requests remediation of a structure that met design standards but failed due to extreme, act of God, circumstances. The engineer must advise the parties that designing beyond the standard of care may be acceptable but could incur unnecessary expense to the owner because the minimum requirements are being exceeded, and should only occur with the owner's concurrence.

• Where engaged to do remediation before being requested to provide expert opinion as to failure/fault:

In all but the simplest cases, a Forensic engineer should avoid recommendations for remediation if an opportunity to design the measure, using current standards, is denied. The engineer should not perform remediation services without thorough investigation of the cause of the failure. The ethical requirements to protect the public health, safety, and welfare demand that any remediation meet the standards of care described earlier. Without full knowledge of the failure, the remediation could be insufficient or extremely conservative.

4.4.17 Relationship with Attorneys, Attorney Ethics

The acrimonious nature of litigation and the close interprofessional relationship of Forensic Engineers with attorneys during investigations of failures require that both subscribe to codes of conduct. Principles for conduct of Forensic Engineers have been the subject of this chapter to this point. Several codes of conduct have been proposed for attorneys who deal with expert witnesses although none has been adopted by any legal groups. The American Bar Association published Standards for Criminal Justice (1980) includes standards for dealing with experts, and emphasizes that attorneys acknowledge and respect the expert's need for impartiality of investigation and independence of thought and opinion.

Feder (1991, p. 235) presents a proposed code of conduct for attorneys who engage in professional associations with expert witnesses. Among other canons, it states that attorneys shall:

- Never proffer an expert with fraudulent credentials,

- Treat expert witnesses with respect,

- Honor expert witnesses' obligations to their codes of ethics, and

- Provide the expert with all the relevant facts and data at their disposal.

As noted, neither this nor any published code has been adopted by attorneys, but these canons provide the essentials of an ethical relationship between and attorney and a forensic expert.

4.4.18 Demeanor

Practically all textbooks and articles on expert testimony stress the importance of demeanor in and out of court. This importance extends from conveying a professional stature to communicating credibility and objectivity. Thompson and Ashcraft (2000) offer the following suggestions for effective demeanor:

- Be yourself.

- Don't be argumentative.

- Don't be demeaning or condescending.

- Be frank.

- Be helpful.

- Be respectful.

- Dress appropriately.

4.5 APPLICATION TO THE HYPOTHETICAL

In the case of the hypothetical bridge failure described in Appendix A, any one or more of the following faults that could have contributed to the failure:

- Missed inspections.

- Undetected corrosion of the structural elements.

- Failure to inspect for scour of soils around pile caps or abutments

- Failure to upgrade the dam (or the bridge) for today's loads and failure standards

- Failure to inspect the bridge immediately after a major event such as an earthquake, flood, or dam failure.

- Failure to note bridge girder or pier damage from floating logs.

- Failure to verify the adequacy of the piles for increased bridge column heights due to a design change.

- A shift in the upstream stream direction, resulting in an adverse angle of attack on the piers, creating pressure loading or scour depths that exceeded the design values.

- Addition of a new deck that exceeded the deck load limits.

- Failure to design the piles for potential flow rates associated with a dam failure.

- Design errors in allowing for earthquake or ice forces.

- Failing to investigate the bearing capacity of the bedrock, failure to extend the borings below the contact line between bedrock and bed sediments.

- Inadequate embedment of piles into bedrock.

- An accident on the bridge that damaged the bridge or dropped a vehicle into the flow that washed up against the deck or piers.

- Scouring of the bed sediment surrounding the piles, resulting in reduction of lateral support.

- Lack of providing or maintaining the scour protection for the piles or abutments.

Other factors may also have contributed, or some of the ones listed may have led to failure of others. Each one of the reasons stated above could have been either the sole reason for the bridge failure or may have contributed to some degree to the failure. The causes, and degree of contribution of each, should be determined based on thorough forensic investigations and appropriate engineering principles and standards.

From an ethics standpoint, an investigation of the mode of failure and time of failure before or after an event is important. Changes in design may or may not have contributed to a failure mechanism. This should be established by examining the revised design and its effects on the structure. For example, changes in pier heights and span clearances could have caused additional moments on the columns. Similarly, if an earthquake or a dam break occurred after the bridge failure, it would be incorrect to conclude that the bridge failed due to those mechanisms.

4.5.1 Historical Causes of Bridge Failure

Statistics show that 60 percent of all U.S. highway bridge failures are due to scour. Actual bridge failures over waterways have been investigated by forensic hydraulic engineers, resulting in findings similar to those listed above. For example, the Schoharie Creek bridge failure in New York in April, 1987, in which 10 people died, was found to be due to failure by the State Thruway Authority to maintain an adequate rock riprap blanket around the piers, resulting in gradual local scour in the glacial till bedrock that eventually exceeded the spread footing depth. Laboratory tests showed that the till eroded continuously at velocities of about 8 fps. Flow rates in the channel were less than this value, but velocities in the horseshoe vortex that surrounds a pier were in the 15 to 20 fps range, resulting in erosion of the unprotected till.

The 1989 Hatchie River bridge failure, in which 8 people died, on Highway 51 in Tennessee was the result of a lateral shift in the channel toward one end that didn't have pilings as deep as those in the main channel. In both cases, the forensic investigation determined that lack of structural redundancy in the design of the bridges contributed to the severity of the failure.

The forensic investigation of the I-5 bridge failure in March 1995 near Sacramento, California, disclosed that the bridge failed due to scour. . Seven lives were lost when 4 cars plunged into the flooding stream after scour washed the footings and caused collapse of the 34-ft long center spans of both bridges. The 122-ft long bridges spanned Arroyo Pasajero Creek, which had an extremely mobile bed and banks. The two bridges were founded on concrete piling, approximately 35 ft in length. The pilings were reinforced for the first 12 ft below the streambed, but had no reinforcement for the remaining 23 ft. It was found that scour removed sufficient lateral constraint to cause structural failure of the piling in the unreinforced length of pile. The bridges were constructed in 1967. Current design standards for scour protection call for a safety factor of 1.0 against such collapse under a fully scoured condition expected in a 500-yr-flood event. A 500-yr event has a 0.2 percent chance of occurring in any year. Scour depths of 30 to 50 ft under such events are common.

4.5.2 Possible Stipulated versus Disputed Facts in the Hypothetical

Factual information in most court cases falls in two categories, undisputed or stipulated facts, and disputed facts. For a failure of this type, both sides in an issue will often stipulate to a sub-set of facts that are not disputed. Examples for this case might include:

- The bridge failed due to either man-caused or natural circumstances.

- A significant flood event occurred.

- Scour around the pile cap occurred during the flood and was a major contributor to the failure.

- The piles were not designed for the increased moments generated due to unsupported column lengths, ice forces, and high water flows after removal of lateral confinement due to scour.

- A hinged structural design would not be used in today's standards even though it was prudent at the time the design was prepared.

The disputed facts that might occur in investigating the hypothetical could encompass any of the following questions:

- Did a change in design produce an inferior product (bridge) compared to the original design?

- Should the bridge have been designed for earthquake forces using a response spectrum?

- Should a hydrologic study have been performed specifically for the bridge, especially because the bridge was located downstream of a dam?

- Should the bridge have been inspected more frequently?

- Did the bridge show any signs of stress between the last inspection in 1997 and the failure in 1999?

- Should the inspections have included examining the soils around the pile foundations for potential scour, especially knowing that a year before there was a dam break with increased water flow and potential for severe scour?

- Did the pile rebar corrode?

- Did an ice jam and breakup contribute to the failure?

- Was the flood magnitude and frequency in excess of permissible design standards?

- Was the flood an "act of God" or did failure of the bridge occur because of a flaw in its ability to withstand the design flood?

- Did the designer prudently apply applicable standards to the design?

It is completely ethical for engineers on each side of a case to seek common ground by stipulating to the obvious, rather than arguing the obvious as often happens.

4.5.3 Standards of Care Relevant to the Hypothetical

The upstream dam was probably designed in accordance with standards in effect in the late 1950s. Those standards may or may not meet today's dam design and seismic design requirements. All medium and large dams have been evaluated for safety since the 1962 Teton dam failure in Idaho. Many dams have required modifications to safely pass floods of this magnitude. The dam break in the hypothetical most likely caused increased flow rates and depths that could have led to excessive scour and eventual failure of the bridge column supports.

The bridge pile foundations should have been designed as being laterally unsupported for the design-flood scour depth. The bridge should have been inspected immediately after the earthquake and dam break. A river bridge is a critical structure and should be designed for earthquake and ice forces. Pile foundations for the bridge should have probably been embedded in bedrock. When downstream of a small to medium size dam, the bridge column supports should have been designed for the maximum flows anticipated under a dam break condition rather than just relying on a government hydrology report.

One ethical issue related to standards of care, for disasters in which nature contributed, is whether the event exceeded the magnitude or frequency of the approved, standard design event. Bridge design standards in the 1950's allowed designs for hydraulic and scour conditions that would exist in a 25- to 50-yr flood. Today's standards call for designs for floods as large as the 100- to 500-yr event. It would be inappropriate in this case for a witness to:

- Fail to describe the standards that were applicable at the time of the design.

- Infer that the design was flawed because the bridge would not withstand a 100-yr event.

- Testify that it was not an act of God if the flood was proven to be considerably greater than the permissible design event.

4.5.4 Ethical Issues Raised by the Hypothetical

Three parties are involved in the hypothetical bridge failure:

- the owner (State),

- the contractor, and

- the bridge design engineer.

As described in the hypothetical, each party chose to hire its own forensic expert. Each stakeholder has different interests, and its expert would probably concentrate their investigation on possible causes of the bridge failure related to possible liabilities of their client. It is advantageous that all three parties sought expert help, because each possible contributing factor should be evaluated completely and separately from a range of viewpoints. This is acceptable provided that each conducts an objective investigation of all reasonable causes, and provided that any allegation made from the experts' investigations is supported by sound and reasonable engineering analysis.

Ethical issues for the hypothetical fall in two categories:

- Whether the state, design engineer or contractor conducted themselves ethically, or

- Ethical breaches that the forensic experts might succumb to in providing the litigation support.

The latter category has been the subject of this chapter. Only the first is discussed here.

4.5.5 Ethical Issues in Each Party's Conduct

The State. The state's responsibility was to review and ultimately approve both the design engineer's design and the design modifications proposed by the contractor. As is the norm, state agencies use in-house professionals or engage independent consultants to conduct these reviews. By resting blame on the hydrologic analyses conducted by the design engineer, or on the design modifications proposed by the contractor, the state appears to be shirking its responsibility of review and approval but probably has no code of conduct regarding its decisions.

Design Engineer. From the data provided, the design engineer appears to have failed to properly follow prudent design guidelines. The design engineer was responsible for independently establishing the hydrologic conditions for the bridge site and apparently did not even check the hydrology provided by the state. Bridge waterway and scour standards require that the designer validate published flows or generate independent, design-level hydrology. If an independent calculation of hydrology for a bridge design is not done, the engineer has the duty to confirm the design flow rates provided by others. Failure to do this reflects imprudent application of design standards.

The designer also neglected to give consideration to failure of the upstream dam, which later breached. The dam may have been declared unsafe, or it failed due to a flood in excess of its design capacity. This appears to have been an oversight by the design engineer, and may not fall in the category of negligence unless the design engineer was cognizant of weaknesses in the upstream dam and did not bring it to the attention of the state.

The data presented was inconclusive as to whether the design engineer acted ethically in reviewing the modifications proposed by the contractor with respect to the bridge framing and pier design.

Contractor. The Contractor was correct in claiming that the proposed design modification was in compliance with the intent of the bridge framing system, but was at fault for not checking the proposed modifications to the piers. This action of excluding review of the design modifications to the piers has ethical ramifications because the consultant apparently declined or was denied the review of these pier modifications.

4.6 DAMAGE DONE BY UNETHICAL CONDUCT

Numerous examples of instances in the forensic expert system could be cited to illustrate the damage being done to the engineering profession (and to the public) by careless or sloppy work or incompetent or unethical engineers and scientists. As an example, an incident reported in an *USA Today* article (April 16, 1997) makes this point. The article describes the findings of the Justice Department report on "shoddy work, lax procedures and questionable practices" by experts in the U.S. FBI Laboratory. According to *USA Today*, the Justice Department findings in regard to investigations and testimony by the "experts" included:

- 'Scientifically flawed testimony' in four cases, including the bombing of the World Trade Center garage.

- 'Inaccurate testimony' by lab examiners in three cases, including the World Trade Center garage case.

- 'Testimony beyond the examiner's expertise' in the World Trade Center garage case and in the investigation of the 1989 crash of Avianca Airlines Flight 203.

- 'Improper preparation of laboratory reports' by technicians in the explosives unit.

- 'Scientifically flawed reports' in the investigation of the Oklahoma City bombing and other cases.

- 'Inadequate record management and retention system' by the lab.

The fallout to the credibility of the expert witness system from this could be substantial. In 1996 alone, the lab analyzed 600,000 pieces of evidence in 20,000 cases. Numerous challenges of these and thousands of other technical cases and convictions could be the result, at considerable injury to the engineering profession, and expense to the taxpayers and impacted parties in the cases.

4.7 REPORTING UNETHICAL CONDUCT

When conduct or services of Forensic Engineers is observed that appear to disregard the safety, health, and welfare of the public, it should be immediately reported. This is a paramount duty, not an option, of all engineers. Serious or repeated breaches of any of the 136 principles of ethical conduct listed in Appendix B should also be reported. Usually, the written allegation should be directed to agencies such as the local building officials, state engineering licensing boards, professional society ethics boards, the court officials, or directly to the owners of the facility or party directly responsible for the public's safety. The duty of investigating and proving the misconduct then passes to the appropriate authority.

The incidence and composition of engineer's reports of possible unethical conduct is unknown. Each state registration or licensing board normally publishes a list of disciplinary activity each year, but no compilation of the national statistics is available. It is likely that the percentage reported is small. Between 1982 and 1990, only three cases involving expert witness services were brought before the Board of Ethical Review of the NSPE. Two related to conflicts of interest and the third dealt with truthfulness and objectivity.

In 1993, the ASFE took over management of the "Expert Witness Deposition Testimony Review Program." This allows attorneys to avail themselves of a panel of experts who will review deposition transcripts and exhibits whenever there is suspicion that the transcribed testimony does not reflect (1) the standard of care, (2) the state of the art, or (3) competency. Submissions may be made only during the pretrial stage of a lawsuit, and a panel of five to six practitioners will be assigned. Scrutiny of negligent misrepresentation of standards of care is the primary focus of the program. Attorneys (or engineers) can initiate the program by contacting ASFE at its Silver Spring, Maryland, headquarters.

Whether directed to ASFE, court officials, the owner, or a registration board, certain items should be disclosed by those filing a report of purported unethical conduct or service. In addition to a complete description of the nature of the unethical conduct or service, the report should include:

- The name, address, and employer of the party alleged to be responsible for the unethical behavior.

- The name, address, and employer of the reporting party.

- The circumstances of the alleged misconduct.

- Any involvement or interest in the case of the reporting party. This should include the scope of services actually performed on the case, the scope of services the reporting party was contracted to perform, and the type of payment terms involved in the contract. If known, these elements should also be reported for the subject engineer.

- The history of the reporting party's involvement with the project.

- Other individuals or parties involved with the project, which may have been directly or indirectly involved with or may also have witnessed the unethical conduct or services being purported.

- A summary, including documents, of the events in the case that led to the allegation of unethical conduct or service. Construction drawings, specifications, exhibits, deposition transcripts, and other records from the case should be provided when relevant to disclosing the unethical conduct.

- A specific description of the allegation, including citations of standards of care, relevant codes of ethics or professional registration documents.

Once unethical conduct or services have been proved, it is the duty of the licensing boards or other authority to take action against the practitioner after the particulars of the case have been presented, heard, and judged. It is also the duty of these entities to recommend, for the particular case of interest, whether additional action is required to restore the safety, health, and welfare of the public.

When reporting apparent unethical conduct or services it is often difficult to maintain the focus of the investigation on the unethical conduct or services rather than the party reporting them. This may be the greatest reason that so few engineers who observe a suspicious incident will report the circumstances. In cases where the reporting party was also a participant in the case, and in particular where the alleged conduct led to a favorable judgment for the perpetrator's client, an air of retribution may exist which should also be evaluated by the authority, but only after thoroughly judging the conduct of the engineer being appraised.

Fear of appearing vindictive should not deter the reporting party. Most states *require* engineers to report incidences of what may be perceived to be unethical behavior. In the absence of state requirements, this duty falls to the self-policing office of the profession and its members. To fail to report does not serve the public's interest and is itself unethical in some jurisdictions.

In conclusion, adherence to the principles and ideals discussed in this chapter by all Forensic Engineers should improve objectivity and mitigate the damage being done to the public and the profession by engineers who engage in advocacy and other unethical practices. Discussions in ASCE's journals of the ethical issues and compiled principles listed in this chapter are encouraged, as are other suggestions for improvements in professional engineering support of litigation.

4.8 ETHICAL STANDARDS FOR PUBLICATIONS

The ASCE has promulgated ethical standards in its *Authors' Guide to Journals, Books, and Reference Publications* (1997), including lists of ethical obligations of authors, reviewers and editors (ASCE 1997). While many of the canons of ethics presented for authors refer to avoidance of fragmentation and submittals of the same material to multiple journals, others relate to objectivity, competence, plagiarism, falsification of data, and the difference between personal and scholarly criticism. It is the position of this set of guidelines that the members of ASCE who prepare written documents for use in court or administrative proceedings should be familiar with and comply with all of ASCE's ethical guidelines, including those listed in the *Authors' Guide to Journals, Books, and Reference Publications* (1997).

4.9 CONCLUSION

In defense of most engineers, a significant portion of the borderline practice seen in the nation's courts may be the result of witnesses being pressured by advocacy-minded attorneys to help them meet their objectives. Engineers tend to have amiable or analytical social styles and may yield to pressures applied by more expressive or driving personalities of attorneys or clients who have a broader view than engineers of the malleability of numbers and technical facts. Regardless, each engineer that defaults to unethical practice is singly accountable, whether or not temptation was offered by an attorney or by the engineer's own conscience.

Professional (hired-gun) witnesses, rather than witness-professionals, harm the profession and create injustices by skillfully and intentionally offering supportive opinions to clients for the price of their time, regardless of the facts and circumstances. Hired guns have been defined in numerous ways. As used in this chapter, they are defined as engineers who are an overall success primarily because they sound good in the courtroom, regardless of the lack of veracity of their testimony. Their chief skill lies in conveying an authoritative tone, even when they are wrong, rather than earning expert status by practicing their trade and testifying with an integrity that compels them to love truth and work harder in seeking truth than most others.

Testimony by high-integrity engineers who find themselves sitting opposite hired guns quite often comes across as less articulate or convincing. It is for this reason that engineers must strive to be both thorough and articulate. Juries are not engineers, and are often more highly impressed by style than substance. This does not imply that the honest engineer must be flamboyant or insincere, or that he or she should be ashamed of not prevailing in presenting the technical facts if honest testimony was given. Any incorrect or unjust decisions that arise in these cases are the fault of the hired guns, not the ethical Forensic Engineers. Unethical practices need to be reported (see recommendations in Section 4.7), but when they occur, they remain unethical regardless of whether someone else observes or reports the conduct.

4.10 REFERENCES

ACEC 1988. "Liability and Litigation Report," *American Consulting Engineers Council*, Vol. 2, No. 6, Nov./Dec.

ACEC 1990. *ACEC Expert Witness Testimony Review Procedure*, Dec. 7.

American Bar Association 1991. *Litigation News*, December.

ASCE, American Society of Civil Engineers, 1927. "Code of Practice," *ASCE Manuals of Engineering Practice, No. 1*, New York, N.Y., Reprinted 1954, 1957, 1960, 1962, Orig. Adoption, Jan. 18.

ASCE, American Society of Civil Engineers, 1961. *Amended Code of Ethics*, ASCE, New York, N.Y., Adopted July 20.

ASCE, American Society of Civil Engineers, 1977. *Amended Code of Ethics*, ASCE, New York, N.Y., Adopted Sept. 25, 1976, Effective January 1, 1977, Amended Oct. 25, 1980.

ASCE, American Society of Civil Engineers, 1989. *Guidelines for Failure Investigation*, Technical Council on Forensic Engineering, New York, NY.

ASCE, American Society of Civil Engineers, 1997. *ASCE Authors' Guide to Journals, Books, and Reference Publications*, ASCE, Reston, VA, May.

ASCE, American Society of Civil Engineers, 2000. *ETHICS: Standards of Professional Conduct for Civil Engineers*, New York, N.Y., April 3.

ASFE, American Society of Foundation Engineers 1987. *EXPERT: A Guide to Forensic Engineering and Service as an Expert Witness*, Assoc. of Soil and Foundation Engineers, Silver Springs, MD.

ASFE, American Society of Foundation Engineers 1996. *Recommended Practices for Design Professionals and Scientists Engaged as Experts for Technical Review of Others' Work and Providing Testimony in Public Forums*, Assoc. of Soil and Foundation Engineers, Silver Springs, MD.

Babitsky, S. 2001. "Becoming and Expert Witness," *Civil Engineering Magazine*, April.

Bachner, J.P. 1988. "Facing Down the Hired Gun," *Journal of Performance of Constructed Facilities*, November.

Budinger, F.C. 1987. "Engineers in Court," *Civil Engineering Magazine*, ASCE, No. 8.

Carper, K. L. 1990. "Ethical Considerations for the Forensic Engineer Serving as an Expert Witness," *Business and Professional Ethics Journal*, Vol. 9, Nos. 1 and 2, Spring-Summer.

Carper, K. L., Ed. 1989. *Forensic Engineering*, Elsevier Science Publishing Co., Inc., New York, NY.

CEC, no date. *Code of Ethics*, Consulting Engineers Council of the United States.

Corley, W.G., and A.G. Davis 2001. "Forensic Engineering Moves Forward," *Civil Engineering*, June.

Dixon, E.J. 1989. *Forensic Engineering – Ethical Practices*, Proceedings, 1989 Annual Conference, ASEE, Washington, DC.

Dolan, T.J. 1973. "So You are Going to Testify as an Expert," *ASTM Standardization News*, ASTM, West Conshohocken, PA, March.

ENR 1987. "The Whole Truth and Nothing But?," *Engineering News Record*, June 4.

Feder, H. A. 1991. *Succeeding as an Expert Witness*, Van Nostrand Reinhold, New York, NY.

Friedlander, M.C. 1990, *A Radical Proposal to Stem the Professional-Liability Crisis*, Consulting/Specifying Engineer, June.

Groden, B.T. 1991. "Is the ASCE Code of Ethics Obsolete in Today's Society?," *Civil Engineering Magazine*, January.

Huber, P. 1991. "Junk Science in the Courtroom," *Forbes Magazine*, July 8.

ICED 1988. *Recommended Practices for Design Professionals Engaged as Experts in the Resolution of Construction Industry Disputes*, Silver Spring, MD.

ICED 1996. *Recommended Practices for Design Professionals and Scientists Engaged as Experts for the Technical Review of Others' Work and Providing Testimony in Public Forums*, Silver Spring, MD.

Johnson, D. 1991. *Ethical Issues in Engineering*, Prentice Hall, USA, 392 p.

Lewis, G.L. 1997. "Objectivity vs. Advocacy in Forensic Engineering," *Proceedings of the First Forensic Engineering Congress*, ASCE, Minneapolis, MN, October.

Macrina, F. 1995. "Scientific Integrity Appropriate? Y: An Introductory Text with Cases," *ASM Press*, USA, 283 p.

NAFE, National Academy of Forensic Engineers 2001. *Guidelines for the P.E. as a Forensic Engineer, the Engineer as an Expert Witness*, January.

NFC, National Forensic Center 1989. *Code of Professional and Ethical Conduct*, (Draft), Lawrenceville, NJ.

NSPE, National Society of Professional Engineers, 1985. "Code of Ethics for Engineers," *NSPE Publication No. 1102*, Revised, March.

NSPE, National Society of Professional Engineers 1985. *Guidelines for the P.E. as a Forensic Engineer*, Alexandria, VA.

NSPE, National Society of Professional Engineers 1989. "Opinions of the Board of Ethical Review," *Volume VI, NSPE*, Alexandria, VA.

NSPE, National Society of Professional Engineers 1997. *NSPE Analysis of Professional Engineer Licensure Laws*, Alexandria, VA.

Peck, J.C. and W.A. Hoch 1988. "Liability and the Standards of Care," *Civil Engineering Magazine*, November.

Postol. L.P. 1987. *A Legal Primer for Expert Witnesses*, For the Defense, Defense Research Institute, Chicago, IL, February.

President's Council on Competitiveness, 1991. "An Item-by-Item Summary of Positions of the Section of Litigation's Counsel to Agenda for Civil Justice Reform in America," *Litigation News, Vol. 17, No. 2.*, Dec.

Priedlander, M, C. 1989. "The Design Professions: Let's Regulate Expert Witnesses," *Civil Engineering, ASCE, No. 4.*

Ratay, R.T. 1997. "Living by the 'Thirteen Commandments' of the Forensic engineer/Expert," *Proceedings of the First Forensic Engineering Congress*, ASCE, Minneapolis, MN, October 5-8.

Smith, T.W. III 1999. ASCE General Counsel, "ASCE ETHICS, Edict, Enforcement, and Education," American Society of Civil Engineers, 1999 Zone Management and Leadership Conference.

Specter, M. M. 1988. "What Does it Take to be a 'Good' Expert Witness?," *ASTM Standardization News*, ASTM, West Conshohocken, PA, Feb.

TCFE 1992. "Minutes" – *Technical Council on Forensic Engineering*, New York, Sept. 12.

TCFE 1989. *Guidelines for Failure Investigation*, Task Committee, Technical Council on Forensic Engineering, ASCE, New York, NY.

Thompson, D.E. and H.W. Ashcraft 2000. Chapter 9 in *Forensic Structural Engineering Handbook*, R.T. Ratay, Editor-in-Chief, McGraw-Hill, New York.

Veitch, T.H. no date. *The Consultant's Guide to Litigation Services: How to be an Expert Witness,* John Wiley & Sons, New Your, NY.

Weingardt, R. 2000. "Let's set our own standard of care." *Structural Engineer*, May.

Worrall, D.G. 1984. "Engineer Experts – The Attorney's Viewpoint," *Journal of the National Academy of Forensic Engineers*, October.

Zickel, L. 1998. Interoffice Memorandum (pers. corr.), Zickel Consulting, Dobbs Ferry, NY.

CHAPTER 5 - THE LEGAL FORUM

If anyone turns a deaf ear to the law, even his prayers are detestable.
- Proverbs 28:9

5.1 INTRODUCTION

The American system of jurisprudence provides for resolution of disputes. Civil wrongs or torts are adjudicated within the court system in a manner similar to criminal justice. Construction disputes may be heard in either the Federal Court or in the local state system.

When two or more parties are unable to settle a dispute, often litigation follows. All parties hire paid advocates to become their voice within the legal system. Plaintiffs file lawsuits and the named parties are then referred to as defendants. Counter claims and cross claims sometimes are subsequently filed against the plaintiff and other defendants respectively.

5.1.1 Chapter Purpose

This chapter briefly describes the American system of jurisprudence. Next, the chapter discusses the relationships of the Forensic engineer with other parties in the system. After these introductory sections, the chapter focuses on admissibility of testimony from a legal perspective and discusses the role of engineers in non-adjudication forums.

It is not the intent of this chapter to serve as a thorough exposition of these topics, but rather to introduce some of the key terminology and outline some of the legal system workings related to Forensic Engineering. Readers seeking more thorough descriptions are referred elsewhere.

5.1.2 Description of the Legal System

Litigation begins with the official filing of the complaint with the appropriate government entity. As the case proceeds each party is entitled to engage in the discovery process where all evidence and testimony any party may offer at trial can be full examined by all other parties. The discovery process is intended to avoid a trial by surprise. By reviewing the total body of evidence, the parties and their attorneys are able to gauge the potential for courtroom success.

In many jurisdictions, construction cases can be designated as complex cases – those involving many complicated issues, requiring extensive discovery and court time. Often interrogatories are passed between parties for aid in the discovery process. An interrogatory is a list of formal questions which one party demands to be answered by another party. For example, it is typical for a defendant to request that a plaintiff outline the proof of his or her case by interrogatory.

The discovery process in a complicated construction case may last for a year or more. During discovery, all witnesses, both fact and expert, are identified. Witnesses are routinely examined in a proceeding known as a deposition. In the deposition, attorneys for the opposing parties examine the witness by extensive questioning. The witness is under oath and a court reporter makes a written record of the proceeding.

Once discovery is completed, the case may be settled or proceeds to trial. Because of the burgeoning caseloads, many jurisdictions have a mandatory settlement process known as mediation. In this forum, the parties to aid in reaching a settlement hire an outside negotiator. This process is "off the record" and no information discovered may be used at trial. This process ends when it becomes clear that a negotiated settlement is not possible. It is not unusual for the mediator to stay involved with the case right up to and sometimes during the trial.

When necessary, some cases are tried in court. Each side presents their evidence and witnesses are examined. The initial testimony by a witness is referred to as direct testimony. Lawyers for opposing parties may then cross-examine the witness. Further questioning may occur and is referred to as redirect and re-cross examination.

Arbitration and mediation are non-judicial forms of litigation. Instead of the judge and jury, an Arbitrator or Mediator may hear the case. A panel of three Arbitrators is usually called a Tribunal. The Arbitrator or Mediator may be a lawyer, engineer, architect, or construction expert. Usually a Tribunal will contain a mix.

Arbitration hearings and mediation proceedings are less formal than that of the courtroom. Stenographic records may or may not be taken. In arbitration, the lawyers present their cases just as they do in the courtroom. The Arbitrator may also pose questions to the witnesses.

Arbitration or mediation are desirable for a number of reasons. The cost to the parties is significantly lower than formal litigation. The discovery process can be streamlined and expedited such that the time to get to an award is reduced. Also, construction industry personnel whom are much more likely to understand the facts than a non-technical jury hear the case. The decision to arbitrate or mediate can be a contractual or legal issue.

There may also be disadvantages to arbitration or mediation in that some of the safeguards of traditional litigation may not be present. Also, disputes involving third parties may require traditional litigation methods.

5.1.3 Forensic Engineers as Experts

The engineer who investigates performance problems and when requested or required offers testimony in a legal forum is referred to as a Forensic engineer. The various aspects of the needs of both the engineering and legal professions are addressed elsewhere in this set of guidelines. Testimony is a vital task that the Forensic engineer must perform.

The court refers to engineers working within the legal system as experts. In the loosest sense, an expert is someone who knows considerably more about a certain topic than the ordinary citizen is. Qualifications necessary to be accepted by the court as an expert vary but generally a Bachelors Degree and a professional license are somewhat of a minimum.

Fact witnesses are people who have knowledge of acts of others or of events. Their testimony is restricted to that of which they have first-hand knowledge. Expert witnesses, after being accepted by the court as such, may offer their opinions on a set of given facts or even on hypothetical situations. Experts may be from any recognized field: accounting, medicine, engineering or even the legal profession.

The legal forum is the domain of the lawyer. Forensic Engineers are relegated to subordinate roles – that of an expert witness. The attorney is similar to the movie director and the Forensic engineer is analogous to the actor. Forensic Engineers are extremely valuable to the legal system and are therefore treated with the credibility they earn in the eyes of their lawyer peers.

5.2 ROLE OF THE FORENSIC ENGINEER AS A WITNESS IN LITIGATION

5.2.1 Deposition Testimony

In depositions, the attorney team for the opposing parties will extensively examine the Forensic engineer. This examination is performed under oath and a written transcript is taken by a qualified court reporter.

Depositions have at least a two-fold purpose: First, they allow the opposing parties an opportunity to fully explore the Forensic engineer's opinions that will be offered at trial. Secondly, the deposition gives the parties a chance to determine the potential credibility of the witness. It is important to note that a jury of non-technical citizens rarely is able to fully comprehend the complicated engineering and construction issues placed before them. Since at trial, the ultimate value of the Forensic engineer's testimony is founded on credibility, his or her demeanor is crucial.

Deposition testimony can be lengthy and usually involves tremendous detail. To the Forensic engineer, this process can be laborious and stressful. Even under these very taxing circumstances, the Forensic engineer should strive to maintain a high degree of professional decorum.

Since the deposition is an examination under oath, the Forensic engineer has but one answer – the truth. The Forensic engineer should take whatever steps needed during the deposition to make sure he or she understands the question being posed. Questions with multiple parts should be addressed one at a time. Questions that involve conflicting circumstances should not be answered. The Forensic engineer should not answer the question the lawyer meant to ask but simply answer what has been asked.

The Forensic engineer can aid this process by asking the lawyer to rephrase or clarify the question. Assuming that the question was mistakenly asked and then answering is a tremendous error. Even though attorneys may not possess an equal technical knowledge, they are extremely intelligent and resourceful. Questions whose meaning may have several interpretations sometimes are used in attempts to discredit the Forensic engineer's trial testimony.

The lawyer for the Forensic engineer's client will undoubtedly have a legal strategy for the case. The Forensic engineer should endeavor to answer deposition questions truthfully but within the imposed strategic goal. In the given hypothetical, the lawyers for the State may want the attorneys for the defendants to be somewhat educated about the facts of the bridge failure. In that case, the State's lawyer may ask the Forensic engineer to answer the defendant's questions fully and perhaps in more detail than is actually being requested.

Clearly there are many strategies which may be employed in a construction lawsuit. Many times, Forensic Engineers may even participate in the development of a particular strategy. This conduct is acceptable as long as the Forensic engineer maintains the objectivity and truthfulness of or her opinions.

5.2.2 Trial Testimony

While the deposition may be lengthy and can focus on intricate detail, trial testimony generally involves key issues and themes. Ideally, construction litigation is a search for the truth. The trial is a test of the legal team's ability to make the complex issues understandable and interesting. The attorney's biggest challenge is to hold the attention of the judge and jury while laying the factual basis for the positions taken in the case.

It is during the trial that the lawyer orchestrates the witnesses and evidence into a plausible and cogent presentation. The sequence of witnesses coupled with tactical considerations will dictate the order of the presentation. Often times, the Forensic engineer is required to be present in the courtroom waiting his or her turn or to listen to others. During those times, the Forensic engineer should remain impassionate to the proceedings and maintain a professional decorum.

While on the witness stand, the Forensic engineer should obviously answer truthfully all questions posed. The answers should be given with the level of detail intrinsic to questions. Since the jury is the finder of fact, the Forensic engineer should respond to questions in a fashion that will be understood by laypersons.

Regardless of legal strategy, there are several things that are inherent to good testimony. The most important is that the Forensic engineer presents a firm belief in his or her opinion based on a thorough understanding of the facts which form the foundation. The most effective way for an adversary to undermine an opinion is to show that the underlying, factual basis is incomplete or inaccurate. The Forensic engineer should avoid speculation; an "I don't know" answer is always better than a guess.

Credibility is the ultimate value of a Forensic engineer's trial testimony. In many cases a jury will not fully comprehend the technical explanation of a Forensic engineer's opinion. The jury will therefore rely on their perception of his or her honesty and expertise of the subject matter. Straightforward answers enhance credibility. The Forensic engineer's reliance on learned treatises as well as his or her own published works are certainly indicative of expertise in the field.

The bane of the Forensic engineer's trial testimony is cross– examination by opposing counsel. It is the duty of the cross – examiner to attempt to undermine the credibility of the Forensic engineer. A resourceful attorney may want to "pull out all the stops" in this endeavor. Cross – examination can be extremely stressful. A Forensic engineer (or any witness) has no control over what questions may be asked. The witness is obligated however only to answer those questions, that he or she feels qualified to answer.

The ultimate success by opposing attorneys is for a Forensic engineer's testimony to actually help the opposing attorney's case. This is fairly rare. It should be pointed out that a lawyer has no obligation to present evidence that hurts his or her case. Invariably there are technical aspects that may be detrimental. The Forensic engineer, therefore, is well served to answer all questions truthfully and in a straight – forward manner.

Forensic Engineers do not win or lose trials; lawyers do. The Forensic engineer, therefore, has no track record of success or failure in litigation. The Forensic engineer's role at trial is to present opinions in a believable and credible way. That is the ultimate determination of success as an expert witness.

5.2.3 Rebuttal Testimony

During trial, the attorney may elect not to call the Forensic engineer as a witness on direct examination. This is a legal decision and is strategy driven. It is not unusual in those circumstances for the Forensic engineer to be asked to serve as a rebuttal witness. Usually this involves listening to other expert testimony and then to be called as a witness.

A Forensic engineer should agree to serve in this capacity only if he or she has complete knowledge of the engineering aspects of the case. To agree to testify on the spur of the moment with a very limited knowledge of the facts is highly unethical.

Testimony on rebuttal follows the same format as direct and cross – examination with one exception. That is, no new facts are allowed to be placed into evidence. The Forensic engineer's rebuttal testimony must be based on information previously offered including testimony by opposing experts.

5.2.4 Arbitration Testimony

A three-judge panel known as a Tribunal usually hears construction cases that go to arbitration. Most often, one or more of the judges have technical expertise pertaining to the contested issues.

Arbitration is a somewhat more relaxed than that of the courtroom with judge and jury. A stenographic record is optional. The court-established rules of evidence are suspended and a set of rules is adopted.

Since the Tribunal is both judge and jury, the Forensic engineer should gear his or her answers to the technical level that is appropriate. Once the direct and cross-examinations are completed, the Tribunal will often engage in further questioning. It is not unusual for the Tribunal to decide a case based on the technical merits. The Forensic engineer should therefore prepare his or her testimony accordingly.

5.2.5 Relationship to The Legal Team

It is the function of the Legal Team to present the best possible case to the court. To do less is to perform a disservice to the client. In development of strategy, often the lawyers will want to discuss the issues with the Forensic engineer. These may include discovery requests, interrogatories, technical issues, and even the approach to cross-examination of opposing experts.

These types of services by the Forensic engineer are ethical only if he or she maintains objectivity. There is nothing wrong with assisting the non-technical attorney in both understanding the complex issues and helping to expose the weaknesses of adversarial experts.

It is, however, considered highly unethical for a Forensic engineer to 'craft' opinions to benefit the case. Conversely, it is acceptable for the Forensic engineer to modify the language needed to communicate opinions such that the non-technical public may more easily understand it. There is a distinction that should not be overlooked. The Forensic engineer should not alter, expand, contract or develop technical opinions to aid a client's position. He or she should endeavor to base opinions on scientific and factual knowledge. The Forensic engineer may ethically substitute more basic or even non-technical language if it assists in understanding.

It is obvious that communication skill is the hallmark of a Forensic engineer as an expert witness. The best scientific mind is wasted in this field of endeavor if the ideas and opinions cannot be made understandable by the legal system and by the general public. This skill is contrasted by stretching the truth or extending an opinion beyond the rational technical justification.

5.2.6 Relationship to Adversaries

Litigation is an adversarial process. Opposing parties will invariably retain both advocates and experts to prosecute their side of the dispute. Forensic Engineers can count on confrontation with both as a case proceeds through discovery to trial.

It is perfectly legitimate for different engineers to draw different conclusions and opinions from the same facts. Different levels of education and experience often guide Forensic Engineers in somewhat different directions. It is unethical for an expert to ignore facts that are contrary to his or her opinions. The Forensic engineer must maintain objectivity throughout the process and is obliged to reevaluate opinions when new facts are discovered.

It should be obvious that the dispute is between the litigating parties – not the experts. That is, the Forensic engineer is an advocate of his or her own opinion and in reality is not a party to the dispute. He or she should strive to establish and maintain a professional demeanor and respectful interface with both the attorneys and experts for the other side.

It is not unusual during depositions or even at trial for an attorney to assail a witness. In depositions, this tactic can be used to test whether the witness will lose his temper or composure. The Forensic engineer is well served to handle this attack with poise and aplomb. The attorney who attempts this tactic in the courtroom is risking his or her own credibility with the jury.

Although the direct affront can be taxing, the Forensic engineer must always refrain from even the appearance of arguing with opposing lawyers. It should be pointed out that by taking part in the debate, the Forensic engineer seemingly becomes part of the dispute – not the desired effect.

In his or her own personal defense, the Forensic engineer certainly is not required to be demeaned. In depositions, the lawyer for the client should request that opposing attorneys stick to the subject matter. More than one deposition has ended with the expert walking out. Naturally, any heated exchanges should be "on the record" since ultimately a judge may have to rule on various aspects.

The Forensic engineer's relationship with opposing experts should be that of equal respect. Simple disagreement of opinions is not justification for unprofessional behavior. Opposing experts are no more part of the dispute than the Forensic engineer is. Cordiality and decorum are the order of the day.

5.3 ROLE OF FORENSIC ENGINEERS IN MEDIATION

As litigation has increased dramatically in this country, the court system has strained to keep pace. In some jurisdictions, the length of time from start to finish may be four years. As a result, different forums of alternate dispute resolution have become popular. The most promising and effective is mediation.

Mediation is a non-adjudication method of dispute resolution. The parties jointly hire a neutral third party to assist in the process. The primary difference between adjudication forums and mediation is that of decision. In adjudication, a third party finds fault and awards accordingly. In mediation, the parties define the resolution.

The mediator's role is central to the process. It is his or her function to hear both sides and without finding fault, assist in structuring a resolution. "Interests" rather than "positions" are defined. The successful mediator is an accomplished negotiator who attempts to help the parties choose an immediate and certain reconciliation over the distant and uncertain adjudication outcome.

The Forensic engineer can and should play an important role in mediation. Settlement of construction cases often centers around the cost of the repair. The Forensic engineer, without attributing blame, can analyze the problem, help both sides to understand what happened and focus the discussion on why. Assuming it is within the scope of his or her engagement and area of expertise, the Forensic engineer can also give parties a preview of what the positional debate at trial will look and sound like when the question of fault is explored.

Mediation is a reconciliation of dispute. The Forensic engineer is not in the dispute nor is he or she an advocate. Mediation is the place to look for solutions and is not a place for ego or unrelenting professional pride.

Often, the issues of a construction case are clear and the culpability easily established. The problem simply boils down to money. That is, the defendants recognize their liability but are without the resources to respond accordingly. Mediation offers the prospect of essentially designing a repair that the defendants can reasonably pay for. The Forensic engineer can aid this effort by identifying what is absolutely essential, what is necessary and what could conceivably wait.

The Forensic engineer called upon to participate in mediation may be of enormous help in advising parties where it is safe and appropriate to make concessions. To provide that input, however, the Forensic engineer must be prepared to consider and help the parties fairly evaluate a full range of alternatives including those which may contradict his or her own opinions.

5.4 ESTABLISHING QUALIFICATIONS OF EXPERT WITNESSES

Qualifications of experts were described in Chapter 2, both as viewed by the profession and by the courts. This section reviews the methods by which courts establish that engineers are qualified to testify regarding the subject matter and give opinions regarding the causes and circumstances of the failure.

5.4.1 Presentation of Qualifications

Engineers are normally asked to provide a professional resume of education and experience to the retaining attorney, court and opposing attorneys ahead of the proceedings or at their commencement. In other cases, the qualifications are established during a deposition, through an interrogatory, or even at the beginning of a trial. Statements of qualifications alone, however, do not assure that the witness will pass the court's tests of his or her "expert" status.

After being sworn and asked to give name and address, both sides will normally cross-examine witnesses regarding qualifications. If an attorney does not want the qualifications presented, he or she may be willing to stipulate that the witness is qualified to testify as an expert and to provide opinions regarding the technical aspects of the case at hand. Without the stipulation, a witness will usually be asked a number of questions by the proponent's attorney regarding his or her qualifications in order to present (and prove) the expert's qualifications. Attempts may be made by the opposing attorney to impeach the witness on the basis that he or she has not kept up on recent developments in the field or has failed to stay informed of changes in practice, but this does not normally affect the acceptance as an expert. Some usual, and a few unusual but precedented, questions asked during presentation of qualifications include:

Usual questions:

• Full name and address,

- current title and duties,

- past and current employment,

- colleges and universities attended,

- degrees earned,

- professional development technical courses and seminars attended,

- refereed publications,

- technical reports written,

- textbooks written,

- knowledge of others writings and books,

- knowledge of standards of practice,

- professional registration (including whether registered in the state of the action), and

- licenses, and whether licensed by examination or not.

Other possible questions:

- Age,

- citizenship,

- salary,

- fees charged for investigation and appearances,

- coursework taken,

- earned versus honorary degrees conferred,

- research projects undertaken,

- teaching and conference or symposium speaking background,

- practical versus academic experience,

- patents,

- participation in professional organizations,

- accomplishments and recognition in the profession, and

- honors and awards conferred.

Often engineers will also be asked to name honors received, or may be asked to provide comments on coursework taken or to recall grade point averages earned. Questioning about specific, related experiences usually follows these general questions. Then, the witness is generally questioned about his or her engagement for the case at hand, followed by questions regarding the investigation conducted in order to develop foundation for the opinions that will be offered. Up to this point, the witness will be able to discuss all the topics listed, but will not be allowed to state opinions about the case until being recognized by the court as a qualified expert. The retaining attorney concludes the questioning about qualifications by offering that the court recognize the engineer as having expert qualifications.

5.4.2 Disclosure of Qualifications

In an effort to avoid the time and expense of depositions, many states (Colorado, for example) have instituted new rules regarding disclosure of witnesses qualifications and opinions. These rules mandate disclosure by each party in litigation of witnesses they plan to call, the documents and exhibits they wish to use, the witnesses' *qualifications,* lists of the witnesses' publications, and statements of the various claims and opinions the witnesses will assert. This tendency toward reform is relevant to qualifications of experts because an engineer who does not provide their qualifications, does not maintain lists of publications or lists of cases in which they have testified, or fails to comply with discovery or provide information on his or her expertise, may be disqualified under the new rules.

Failure to fully disclose previous engagements can be detrimental, as illustrated in an incident in the 1998 trial of Terry Nichols for the Oklahoma City bombing. An expert testifying for the prosecution about drill bit markings in a padlock failed to include in his resume the fact that he had been a ballistics witness for the government in the Ruby Ridge investigations where both the FBI and ATF were investigated. The resume had been submitted, but excluded this element. It may not have been relevant because ballistics and padlock markings are not linked, but it was challenged on the basis that the expert might be a hired gun for the agencies. This nearly led to his disqualification, and its intentional exclusion probably discredited his testimony somewhat.

5.4.3 Challenges to Qualifications

After being offered as an expert by the proponent attorney, the opposing attorney may *voir dire* the witness to probe more deeply into some of the statements made, and will often challenge the witness' qualifications, either during the *voir dire* or on later cross-examination, on one or more of several grounds:

- That the engineer's experience is not strong in the particular field of concern.

- That the engineer's training in the particular field of concern is not adequate.

- That the engineer's training does not encompass all the subjects to be covered.

- That the engineer's age proves that he or she does not have the experience needed to qualify as an expert.

- That the engineer has a connection with the case and may have a bias.

- That the engineer has overstated his or her qualifications.

- That the engineer is a "professional witness" whose opinion may be suspect.

- That the engineer only works for one particular type of client.

- That the engineer has only worked as a plaintiff's (or defendant's) witness.

- That the engineer's conduct has been reported to licensing authorities in the past.

- That the engineer may not have designed similar or numerous constructed facilities (this criticism is often directed at university professors)

- That the engineer has publications or writings that affect his or her ability to be objective in the case at hand.

- That the engineer has writings or teachings that contradict his or her opinions in the case at hand.

- That the engineer does not have sufficient knowledge of the applicable standards.

- That the engineer is not knowledgeable on recent developments in the field.

- That the engineer has failed to stay informed of changes in the practice.

Because no testimony on the subject area has been given, the last five questions may not occur during *voir dire*, but could be asked later during cross-examination. Questioning on past failures to pay taxes, failure to pay child support, failure to repay student loans, or misdemeanor or felony convictions are disquieting but may be allowed in some jurisdictions. Courts may permit such questioning during *voir dire* because some state laws allow the professional registration board to take disciplinary action for these infractions (this is discussed in more detail later). Attorneys will often go beyond valid pursuits into various attempts to unduly discredit a witness. The witness may be asked questions such as, "How much does your opinion cost?" The engineer must carefully listen to the question to avoid getting caught in such obvious attempts to shade or discredit his or her qualifications.

5.5 ADMISSIBILITY OF TESTIMONY BY FORENSIC ENGINEERS

5.5.1 Frye

In 1923 the Court of Appeals for the District of Columbia affirmed a lower court's decision, holding that in order for expert testimony to be admissible, "the thing from which the deduction is made must be sufficiently established to have gained general acceptance in the particular field in which it belongs." Frye was a criminal case in which the defendant sought to introduce expert testimony concerning the results of a polygraph examination. The court held that the scientific community did not generally accept the polygraph examination. As a result, the polygraph examination did not rise to the level of information upon which an expert could render an opinion.

Although inconspicuous in its beginning, as the court cited no authority to support its decision, Frye is an example of "common law" springing into reality. Frye became the "common law" relied upon by the federal courts concerning whether specific information was a proper basis for expert testimony. Seeking to control the scientific communities of which their products were a part, "Corporate America" held fast to the Frye doctrine in order to keep plaintiffs' legitimate evidence out of court.

Even if a party's expert was highly qualified, under Frye the party seeking to introduce the expert testimony had to prove that the information the expert relied upon in formulating his or her opinion was something collectively agreed upon by the scientific community. The Frye test became the threshold requirement for the admissibility of expert testimony in federal courts.

5.5.2 Rule 702 - Testimony by Experts

This rule states that if scientific, technical, or other specialized knowledge will assist the trier of fact to understand the evidence or to determine a fact in issue, a witness qualified as an expert by knowledge, skill, experience, training, or education, may testify thereto in the form of an opinion or otherwise.

5.5.3 Rule 703 - Reasonableness

Federal Rule of Evidence 703 addresses the bases of opinion testimony by experts and actually relaxes the requirement that the witness must have personal knowledge of the matter about which he or she is to testify, providing:

The facts or data in the particular case upon which an expert bases an opinion or inference may be those perceived by or made known to the expert at or before the hearing. If of a type reasonably relied upon by experts in the particular field in forming opinions or inferences upon the subject, the facts or data need not be admissible in evidence.

The purpose of this rule is to assure that an expert's opinion is supported by reliable information. Rule 703 allows experts to base their opinions on data that is reliable, yet inadmissible, thus broadening the basis for the opinion. However, some circuits hold that Rule 703 does not apply where the expert's opinions are based on evidence that is otherwise admissible.

Of course, an issue exists as to the meaning of "reasonably relied upon" and how much inquiry the court can make as to the data's reliability. Rule 702 actually requires the courts to evaluate the basis for an expert's testimony. Some of the more liberal courts simply presume that the expert testimony is reliable unless opposing counsel submits contrary evidence. Even if the data supporting an expert's opinion is determined to be reasonably relied upon, this does not mean that the data "itself is independently admissible in evidence." If inadmissible evidence is admitted in order to explain the expert's opinion, opposing counsel can request that the jury be instructed that the evidence is only to be considered "solely as a basis for the expert opinion and not as substantive evidence."

5.5.4 Daubert

With the adoption of the Federal Rules of Evidence in 1975, various federal courts began interpreting Frye in light of Federal Rule 702, Testimony by Experts. As Frye is inconsistent with the "liberal thrust" of the Federal Rules of Evidence, different circuits developed different interpretations as to the admissibility of expert testimony under Rule 702. In the wake of the confusion, the United States Supreme Court granted certiorari in the case of Daubert v. Merrell Dow Pharmaceuticals, Inc. to consider whether the Federal Rules of Evidence superseded the Frye "general acceptance" test.

In Daubert, two mothers alleged that the drug Bendectin, which they ingested while pregnant, caused limb reduction birth defects in their children. The defense presented an affidavit of their expert purporting that use of Bendectin was not causally connected to human birth defects. In response, the plaintiffs presented eight affidavits from experts delineating the causal connection between the ingestion of the drug and birth defects. Applying the Frye test, the trial judge found that even though the plaintiffs' experts may be highly knowledgeable, the subject matter of their testimony was not generally accepted in the scientific community. The Ninth Circuit Court of Appeals affirmed. Vacating and remanding the case, the United States Supreme Court held that the Federal Rules of Evidence superseded the Frye "general acceptance" test. The court reasoned that "nothing in the test of this Rule establishes 'general acceptance' as an absolute prerequisite to admissibility... .The drafting history makes no mention of Frye, and a rigid 'general acceptance' requirement would be at odds with the 'liberal thrust' of the Federal Rules and their 'general approach of relaxing the traditional barriers to 'opinion' testimony.'"

After answering the very narrow question at issue, the Supreme Court continued to analyze Rule 702 in a somewhat advisory fashion and developed a two-prong test that must be met for admitting expert scientific testimony in a federal trial court. The first prong requires that the expert testimony be based on scientific knowledge, while the second prong mandates that the testimony help the trier of fact in understanding the evidence or determining a fact in issue.

Next, to further assist the trial judge, the Court set forth four nonexclusive factors that federal judges ought to consider in carrying out their "gate keeper" function under Rule 702; (1) whether the theory or technique has been tested; (2) whether the theory or technique has been subjected to peer review and publication; (3) the known or potential rate of error, and existence and maintenance of standards controlling the technique's operation; and (4) the "general acceptance" of the scientific theory.

The above factors are merely suggestions of the Supreme Court for consideration by the federal trial judge. However, in going beyond the issue presented, the Supreme Court "resurrected" the "general acceptance" test employed in Frye, once again making it difficult for attorneys to bring cutting edge science into the courtroom.

Furthermore, on remand, the Ninth Circuit Court of Appeals, having previously affirmed Frye, did not believe the Supreme Court's suggested factors for admissibility were sufficient and added an additional factor that the Ninth Circuit considered "very significant" – whether the expert testimony is based on research conducted independent of litigation or expressly for the purpose of litigation. This factor imposes a tremendous burden on the plaintiff in that the testimony is considered suspect if the research was conducted in anticipation of litigation.

Under Rule 702, the trial judge serves as the "gate keeper" charged with the duty of warranting that the expert's testimony "both rests on a reliable foundation and is relevant to the task at hand." In short, after the trial judge qualifies a witness as an expert in a particular field, the Daubert case instructs the trial judge to take off the judicial black robe and step into the white coat of a scientist to determine whether the expert's testimony is reliable.

5.5.5 Kumho Tire Co. Ltd., et al., versus Patrick Charmichael, et al

In March 1999 the U.S. Supreme Court, in Kumho Tire Co. Ltd., et al., versus Patrick Charmichael, et al., settled the issue of whether Daubert applies to engineers. The court ruled that judges can apply the same four tests previously targeted at medical and scientific professionals to expert testimony by engineers. Based on this finding, an engineer's testimony may legally be deemed inadmissible if his or her opinion differs with standards of practice or is not consistent with generally accepted methods.

5.5.6 South Carolina Law (One State's Interpretation)

A quick review of South Carolina law reveals that South Carolina courts apply neither Frye nor Daubert in determining the admissibility of expert scientific testimony. Instead, South Carolina employs a more liberal standard, allowing the jury to serve as the proverbial "gate keeper" in determining the reliability of expert testimony. The adoption of the New South Carolina Rules of Evidence probably will not affect this standard because among other reasons, even though SCRE 702 is identical to FRE 702, SCRE 702 mirrors South Carolina's former provisions regarding admissibility of expert testimony.

South Carolina courts specifically refused to adopt the Frye test in State v. Jones. In Jones, a rape case, the prosecution introduced "bite-mark" expert opinion evidence to show that a photograph of a bite mark found on the victim matched impressions of the defendant's teeth. The trial judge allowed a pathology photographer to testify that the photograph accurately depicted the size of the teeth, and a forensic odontologist to testify that the impressions of the defendant's teeth matched the photograph.

The current law of South Carolina with respect to the admissibility of expert testimony began with the Jones opinion. Affirming the trial court's decision the South Carolina Supreme Court held in very broad language that the admissibility of the expert scientific testimony depends on the "degree to which the trier of fact must accept, on faith, scientific hypotheses not capable of proof or disproof in court and not even generally accepted outside of the courtroom." The "faith" standard established here is not very scientific, but continues to serve as the basis for admissibility of expert scientific testimony. According to the Jones opinion, in applying the "faith" standard, a court does not have to sacrifice its good judgment and common sense in determining the admissibility of expert scientific testimony in South Carolina.

The next relevant South Carolina case in the area of expert scientific testimony admissibility is State v. Myers. Myers is a blood spatter case in which the prosecution wanted to admit the expert opinion testimony of an EMS technician on blood spatter interpretation. This EMS technician had undergone training that included information on blood spatter interpretation, but the trial court refused to allow the witness to testify as an expert for lack of qualification.

On appeal, the South Carolina Supreme Court held that the issue of who is qualified to testify as an expert witness is within the sound judgment of the trial judge and that the burden of proving an expert's qualification falls on the party offering the testimony.   However, the Court went on to say that the failure to show the qualification of a witness as an expert goes to the weight of the testimony, not the testimony's admissibility. The Court concluded that the trial judge's refusal to allow the witness to testify as a blood spatter interpretation expert was prejudicial abuse of discretion. This opinion establishes a fairly low threshold as to who can render an opinion as an expert witness, placing the burden on the jury to determine the reliability and credibility of the expert testimony.

The South Carolina Supreme Court also decided State v. Ford, the first DNA evidence case in South Carolina, in 1990.   The issue before the Court was the admissibility of DNA Print Identification evidence. In holding that such evidence was admissible under the standard set forth in Jones, the Court stated that South Carolina has never adopted the Frye test and "has employed a less restrictive standard in regard to the admissibility of scientific evidence."   Apparently, the Court was not applying the law of South Carolina in noting that DNA Print Identification also met the Frye test, but was merely trying to cover itself on all grounds.

In State v. Dinkins, one of the most recent cases addressing the admissibility of expert scientific testimony, the Supreme Court determined that DNA population frequency statistics were admissible under Rule 24(a) SCRCrimP, pointing out that Rule 24(a) was identical to Federal Rule of Evidence 702, and that Frye was no longer the standard of admissibility. The defense alleged that the expert testimony regarding these probability statistics was prejudicial because the jury may perceive the test as infallible.   The court, referring to Daubert, pointed out that vigorous cross-examination and the introduction of contrary evidence were sufficient tools provided the defense to attack "shaky but admissible evidence."   Although the Court referred to Daubert in part, the Court did not adopt the strict Daubert standard.   Instead the Court refers to the standards set forth in Jones, supra and Ford, supra.

5.6 ROLE OF THE FORENSIC ENGINEER OUTSIDE OF EXPERT TESTIMONY

5.6.1 Consultant to the Legal Team

In many construction cases, lawyers have the need for expert advice but not necessarily for testimony purposes. Often this role is that of educator about the construction process. Occasionally, the lawyer is in need of knowledge that will help him or her identify potential defendants.

While most attorneys will develop their own legal strategy, they lack sufficient technical knowledge to fully understand the various merits. Consulting with lawyers on strategy is a perfectly acceptable undertaking by a Forensic engineer. What experts are needed, what discovery should be undertaken, how the courtroom case should be structured or even how to attack an opposing expert's credibility are all examples of questions the Forensic engineer can ethically address.

The Forensic engineer should not endeavor to "play lawyer". The Forensic engineer should restrict his or her purview to a strict technical sense. In this role, the Forensic engineer should portray himself or herself as an idea source.

5.6.2 As a Defense Expert

There are always at least two sides in a dispute. There will, therefore, be the need for defendants to have their own experts. The role of the Forensic engineer as a defense expert is exactly the same as for the plaintiff side. He or she has no less or no greater duty. It can be said that the opinions developed are based upon fact and are independent of client need.

Engineers do tend to criticize other's work. This tendency may manifest itself to a greater degree when one reviews the report of a plaintiff's expert. The Forensic engineer should strive to objectively investigate the facts and independently develop his or her own opinions. Although the opportunity to 'craft' workable opinions for direct client benefit presents itself to a defense expert, the Forensic engineer should resist all pressures to do so.

5.6.3 As an Insurance Industry Expert

Insurance companies ultimately pay many construction claims. As such, the insurance industry has   precisely the same need as other litigation parties – to adequately assess their potential liability. The Forensic engineer who finds himself or herself working continually for the insurance industry should make concerted efforts at maintaining objectivity.

Seldom is a claim filed in a construction case that is completely without technical merit. All Forensic Engineers, including those working for the insurance industry, should overtly agree where applicable with the technical points that have been put forth by others.

Many times in a construction case, liability is clear. The defendant strategy, therefore, goes toward reducing the dollar amount. The investigation of a performance deficiency usually identifies adjunct violations of standards and codes somewhat unrelated to the initial problem. It is highly unethical for Forensic Engineers to testify, for example, that even though a code violation (negligence *per se* in most jurisdictions) exists that it is non-consequential or irrelevant.

5.6.4 As a Claims Consultant

With construction litigation burgeoning, a somewhat new role for construction industry personnel has emerged. The role is that of the claims consultant. The claims consultant's function is to fully assess the merits and issues, estimate the value of the claim and advise parties as to the potential for success.

The claims consultant can come from almost any allied field of construction or litigation; i.e., engineering, contracting, legal or even accounting. In this role the claims consultant is very much an advocate.

Forensic Engineers who function as claims consultants have the same responsibilities with regard to objectivity of testimony as all Forensic Engineers. In order to maintain credibility as a witness, the Forensic engineer should probably avoid testifying in cases where he or she is serving as a claims consultant.

CHAPTER 6 - BUSINESS CONSIDERATIONS

The Lord abhors dishonest scales, but accurate weights are His delight.

- Proverbs 11:1

6.1 INTRODUCTION

The preceding chapters have dealt with the legal, ethical and technical considerations in Forensic Engineering practice. As is true for all professions, Forensic Engineers are basically businesspersons and must succeed financially.

6.1.1 Chapter Purpose

The practice of the Forensic Engineer must fall within ethical business requirements. This chapter discusses business practices that appropriately meet the manifold requirements of the engineering profession including marketing, pricing and compensation, assignments, insurance, and liability.

6.2 MARKETING FORENSIC SERVICES

6.2.1 Types of Marketing

As with any other professional service, the Forensic engineer wants to have his or her name and reputation brought to the attention of clients in the best light, and to be considered when an appropriate opportunity develops. The Forensic engineer wants to know that the potential client is properly aware of his or her area of expertise and that he or she will be considered when the need arises for Forensic Engineering services. The Forensic engineer has to be selective in the method of marketing services in order to promote a professional image and not to appear as a "hired gun" awaiting a call to do battle for whoever is willing to pay. Here are some guidelines that can help you, as a practicing Forensic engineer, succeed financially while maintaining high ethical standards.

Statements made in marketing literature can become a pitfall if not carefully chosen. Statements can be used against the advertiser as can any statement published or given at a deposition or trial. If a Forensic engineer states that he or she specializes in plaintiff or defendant's cases, this might be interpreted as creating a bias. Statements such as "we are prepared to investigate and scope out any position" can be seen as a willingness to create technical positions where none fairly exist. Marketing statements should be truthful and accurate. There is always an amount of self-praise in any advertising, but this has to be done in a professional manner. Remember, whatever statement is made can be read back to you in court. These various types of marketing are described below.

6.2.1.1 Advertising

Advertising can occur in several formats. Most advertising appears in print or on television or even on billboards. Some law firms are very aggressive advertisers with television, newspaper and billboard advertisements that promote personal injury and worker's compensation case representation. Medical doctors (particularly cosmetic surgeons) and orthodontists are also advertisers in the popular media.

ASCE members are restrained by ethical considerations in the advertising statements they can make. Canon #3 states: "Engineers shall issue public statements only in an objective and truthful manner." Canon #5 states: "Engineers shall build their professional reputation on the merit of their services and shall not compete unfairly with others." Hyperbole and statements deriding competitors are unsuitable in advertising professional services.

Forensic Engineers, therefore, have to be more conservative in their media appearances. Because of the narrow client audience, mass media advertising is not cost effective and could be problematic in the image it would convey.

The common form of advertising for forensic services is in the print media. Business card type listings placed in specific interest journals such as insurance claims adjuster's journals, legal journals and contractor's magazines are a common form of advertising. These listings describe the specialty of the practitioner and provide an address and telephone contact. Such listings are limited in their ability to distinguish one expert from another, but these listings at least put your name before potential clients.

There are several clearing-houses that provide lists of Forensic Engineers. For a fee, which is dependent on various factors including the size of the listing, you can be listed in a book or on a CD that compiles forensic practitioners according to their specialty. These listings are then marketed by the listing service to appropriate potential clients. There are mixed opinions on the value of these listings. Some engineers and their clients have found them to be useful. Others have expressed disappointment with the results.

Brochures, flyers and pamphlets are probably the most common form of print advertising used by Forensic Engineers. These materials allow more space to describe accomplishments, scope of expertise of the individual or firm and the educational and professional background of individual investigators. Brochures can create an image through graphics and color. They can become quite expensive, however. A balance has to be reached between presenting your credentials fairly and engaging in needlessly large costs for the preparation and printing of brochures. The important point is to be able to get your message across. Increasing the cost of the brochure may not increase the effectiveness of the message.

Printed materials are distributed in several ways. One way is to mail service brochures to potential clients using mailing lists. The mailing lists may be purchased from professional organizations or from firms that specialize in the production of targeted mailing lists. This is rather a shotgun type of approach since commercial mailing lists include many names that would not have an interest in your services. Better mailing lists are those generated from selected lists. Parties that have responded to advertisements and listings in the past and have requested additional information are one source. Former and current clients are another source of names. Large law firms and insurance companies consist of many contacts. Just because you are known by several attorneys or adjusters in an office, it does not assure your being a familiar name to others in the same office. It is a good idea to obtain the names of as many potential users of your services as possible and to see that your brochure and flyers are placed in their hands.

Printed materials can also be distributed by hand delivery. A representative of your company can make a personal call on a potential client and leave materials after a short visit in which he or she describes your services. This works for larger firms which can afford the expense of a marketer to do this. A smaller practitioner can do this also but time is taken from billable hours. The amount of available time you can spend and the amount of coverage is limited.

Printed materials can also be submitted with proposals. A proposal for a specific project can be supplemented with a brochure that covers items not specifically requested but which can be of interest to the reviewers. The brochure will convey your message in the specific for you desire.

One other source of mailing addresses for brochures is in response to a direct inquiry for information about you and your services. This inquiry could have come as a result of an advertisement you placed, a flyer you mailed out, a referral by someone who knows of you, or from someone you met at a meeting or conference,

6.2.1.2 Personal Contact

"Cold call." The word sends shivers up and down your spine. You must get access to someone you have never met before and then interest them in your services. Unless you are a born salesman, this is not the type of contact you enjoy.. Happily, such contacts are less commonly used by practitioners because they are not the most efficient use of their time. Cold calling may not assist in providing the image of a successful professional if it appears that you have the time available to make these calls. Larger organizations with a marketing staff are more apt to engage in this form of personal contact.

A more efficient type of personal contact would be a call on a "qualified" prospect. This would be a person or organization that you know has need for forensic services that you offer. You have identified the person or persons that have the authority to engage Forensic Engineering consultants. You meet with the person or persons either on a one-to-one basis or as a group. You know that you are presenting your message to a potential client. The difficult part of this type of direct contact is finding the right person. This may be the result of cold calls by you or by others. This may result from telephone calls to the organization in which you are transferred from office to office until you reach the right people. Newspaper or magazine articles or trade journals may be other sources of names of purchasers of forensic services. Qualified prospects are obtained from referrals, discussed further and from group presentations you may have made, also discussed below.

6.2.1.3 Referrals

Referrals are a good source of marketing leads and new business. A satisfied client who knows of someone with a possible need for your services can recommend you to them. You have now have two points in your favor: meeting with a potential purchaser of your services and having a favorable recommendation from someone known to the potential client. It is therefore very important that you try to obtain referrals from clients. Ask for referrals. Would your client speak to other attorneys in the firm about you and pass your resume around? Would your client introduce you to an associate? Would your client join you for lunch with a referral? Are there insurance adjusters in other branches that utilize forensic consultants? You did a good job for this client. Would the client help you by referring you to someone else he or she knows? A client who knows you and likes your work is your best salesperson in the organization.

6.2.1.4 Repeat Business

Repeat business is very satisfying and is the easiest way of marketing. By doing your job well you market yourself for the next case. Do not be afraid to ask for more work. Be sure that the client is fully aware of your capabilities and those of your firm. Just because you are working on one matter involving a portion of your expertise, it does not mean that the client is fully aware of your capabilities. This may be especially true when your marketing staff did the marketing for the present case. The client knows you are experienced in building failure analysis but may be unaware of your expertise in construction scheduling. Get to know about your client and his or her organization so that you can ask the right questions to generate more work. What type of practice does the attorney's firm specialize in? Who handles larger property claims for your insurance company client?

6.2.1.5 Panels and Other Presentations

Another effective marketing strategy is one involving educational or informative presentations to prospective clients. The programs may be free programs offered at the client's offices. An example would be a presentation to insurance adjusters at a regional claims office dealing with how an engineer would investigate storm or earthquake damage to structures. The presentation would focus on how the engineer would be of assistance to the property adjuster in adjusting such claims. The property adjuster would benefit from the exposure to engineering thought and might be favorably impressed so that you would be called upon when engineering services are judged to be required. The marketing required for these programs involves finding the supervisory person who can arrange for such programs and who can select the appropriate staff who will be in attendance.

Paid seminars that are offered at off-site locations for a larger audience are another type of presentation. These may be one or more days in length. They are typically offered to an audience drawn from many companies and are intended to break even or make a profit. There also may be several speakers from one or more firms offered services to the potential clients. Construction delay claim seminars are an example of such presentations. These have been offered on a one to three day basis with speakers representing the legal, engineering and financial areas of expertise. The audience is drawn from the suppliers and purchasers of construction services. The main purpose is to educate the audience on the handling of routine claim situations. It is hoped that when major claims situations arise, the seminar leaders, who have impressed the audience with their expertise, will be called upon for consultation or major assistance. This marketing strategy has been quite effective for many consultants.

Participation in a program with other experts for a client's professional association is another good forum. Professional groups such as the American Bar Association have various interest groups that utilize experts and attorneys to develop presentations at conventions and seminars. Participation on such panels indicates acceptance by the sponsoring group and adds to your credibility.

6.2.2 Scope of Marketing

6.2.2.1 Presentation of Qualifications

In preparing a marketing plan or strategy, you want to focus on what information you plan to disseminate about yourself and your organization, what type of work you are prepared to perform, and who you want to target as your clients.

The first item would be presenting the qualifications of your organization. Because Forensic Engineering practice depends on the quality of the individual experts involved, knowledgeable clients or potential clients will examine the qualifications of individual experts sometimes more often than the reputation of the organization. Many users of Forensic Engineering services find an expert they feel to be reliable and stick with that person whether he or she works for Company A or Company B or works as a sole practitioner. Even in large, well-respected Forensic Engineering firms, the credentials of particular experts are scrutinized. A large organization can use its collective reputation to develop the assignment, but the successful performance of the work depends on the quality of the individual. This is especially true when courtroom testimony is involved. A "name" in the Forensic Engineering field means little to a jury. What is said and how, is the basis for the rating given by the jury.

Typically in marketing materials, the collective experience of the organization is presented; past investigations, venues at which presentations were made, client lists and hardware and software capabilities. Next would follow individual qualifications in the areas of expertise of the firm. The statements of qualifications could include:

- Educational background in the field of expertise

- Practice in the field

- Licensure, memberships in professional societies or associations

- Membership on code or standards writing bodies or agencies

- Publications or presentations in the field of expertise

- Past cases in the subject area

- Testimony at arbitration or trial

- Courts in which the person was admitted as an expert

6.2.2.2 Multidisciplinary Assignments

The advantage of a group practice in Forensic Engineering over a sole practitioner is in the situation where assignments require expertise in several disciplines. Most major investigations and even many moderate investigations require the use of several specialists. Building cladding problems might require the collaboration of an architect, a structural engineer and a materials engineer. Geotechnical and structural engineers work together when settlements and collapses occur. Investigations of fire damaged buildings utilize environmental engineers, metallurgists, structural engineers, architects, fire protection engineers, and other disciplines. Few Forensic Engineering firms, including the largest, have the full range of experts on staff. Project staff can be developed by having outside consultants on an as-needed basis or by subcontracting or joint venturing with a specialty firm such as a metallurgical laboratory. For marketing purposes this means that you have to have an affiliation plan worked out in advance with people or firms that you have confidence in. You can then proceed to market your firm as being able to handle multidisciplinary assignments of various types. Alternately, should you be asked if you are capable of responding to a particular requirement, your action plan has been prepared and you can go forward. Sometimes a client requests something you have not thought of investigating. Having contacts in other disciplines broadens your ability to respond or to recommend help to a client.

6.2.2.3 Demonstrations

When you are at the point of making a personal presentation, demonstrations of your organization's capabilities can be extremely useful. The presentation might be for an interest group such as attorneys or claims adjusters or might be for an interview for service on a specific assignment. The person or persons you are meeting are interested in demonstrations of your work product. The difficulty in using work product is confidentiality. Using an active case can violate the confidentiality agreement between you and the client. If it is possible to use materials from an active case, (edited to hide actual participants) permission must be obtained from the client.

Old cases that have been closed or settled are better for demonstration of work product, but even then, client permission should be obtained. An out of court settlement may have involved confidentiality agreements between the parties. Other parties to the case may still be actively involved. An example might be an arbitration hearing that has concluded and judgment rendered. One of the parties refuses to pay the award. Though the arbitration itself is over, the parties are still continuing their adversarial relationships and dissemination of any details of the arbitration is to be avoided.

Once the confidentiality issue is addressed, you want to demonstrate the quality of your work product. This may take the form of:

a.   A demonstration of the clarity of your analysis and written presentation. These could be demonstrated through sample reports that show typical format and style of writing. Is your writing clear and to the point or is it filled with technical jargon?

b. A demonstration of your analytical capabilities in the form of hardware and software.

c. Demonstration of your firm's computer graphics capabilities. Demonstration of computer animation and digitization of photographs are an example.

d. A presentation from a compact disc (CD) makes a very impressive demonstration of both the materials you want to present and the means by which they are presented. Having the materials on a CD allows you to tailor your presentation to the audience.

Web sites on the Internet also offer the ability to provide access to demonstrations of your capabilities. Animated as well as static displays are possible.

6.2.2.4 Insurance Companies Versus Attorneys

The client base that a Forensic Engineering firm develops may be diverse or may be restricted. This may come about by choice or may result from the nature of the marketplace with which the firm's principals are working. Generally speaking, clients for Forensic Engineering services can be either the direct principals who require the services or indirect parties such as attorneys who represent principals or the insurance companies for the principals. Who is in the market for forensic services at any time depends on the nature of the loss, the amount of the loss and whether there is insurance coverage.

Typically, larger businesses and public agencies are examples of principals that would retain forensic consultants. They are involved with higher dollar projects, have greater exposure from losses and have the financial resources to engage consultants. Particularly in matters that are not covered by their insurance or for which they are self-insured, they have to find required experts and manage the forensic investigation on their own. Examples of forensic investigations for principals are design and construction defect claims brought by building owners against design professionals and construction contractors. Construction delays or performance problems on public works projects can lead to government agencies retaining Forensic Engineers. On the other side of the coin, errors and omissions claims may bring insurance coverage into play and it will be the insurance companies that hire the Forensic Engineering experts for defense.

If the claims against targeted defendants are likely headed into the legal arena, attorneys representing the insurance carrier become involved early in the process. Whether in-house counsel or outside counsel is retained depends on the nature of the matter and the policy of the insurance company.

For property losses insurance companies will retain forensic consultants directly, since they are initially needed for their technical expertise. Only if settlement or coverage issues develop will the matter be drawn into the legal forum.

The question is then, to whom should you market? The answer depends on the size of your organization, the size and nature of your marketing effort, the nature of your practice, your geographical location and your past experience. Do you want to create niche marketing? What has your clientele been in the past? Some forensic firms specialize in the insurance company marketplace. They offer a wide range of engineering and technical consultants through the use of outside consultants and strategic affiliations. They are prepared to respond to the diversity of technical needs arising from insurance property and liability losses. They directly market to claims adjusters and insurance executives. Other firms market to owners and developers. Their expertise has focused on architectural issues such as leaking roofs and facades.

Support requirements, need for professional liability insurance and cash flow also dictate client targeting. Small plaintiffs in accident cases may represent financial risk for their experts unless significant retainers are obtained. Public agencies and insurance companies have adequate funds but pay requests must be submitted according to procedure and payment may be slow. Larger cases and larger clients offer more opportunities but may require more cash output up front for travel and expenses.

If you are hired by an attorney, you should know if he or she is guaranteeing your fees or is just passing your bills on to the ultimate client, the insurance company, corporation or individual.

6.2.2.5 Advocacy in Marketing

A final word on marketing. When preparing your marketing message you want to impress potential clients with your capabilities and demonstrate to them how you can assist them with their Forensic Engineering needs. Caution has to be exhibited here because of the temptation to appear to be too much of an advocate for the client. The purpose of an expert is to perform a fair analysis in order to assist the judge or jury in understanding technical matters not understandable to lay people. Performing investigations or analyses where the sole purpose is to bolster the client's position places the expert in the role of an advocate. Attorneys are advocates. Forensic Engineers are expected to reach rational conclusions based on thorough engineering analyses.

Your marketing should indicate your need to conduct an appropriate engineering investigation in order to reach your conclusions. The purpose in retaining you is to utilize your expertise. The results of your investigation are not known beforehand. There are instances where marketers of Forensic Engineering services (engineers in their own right) have sold their services indicating beforehand that a multi-million dollar recovery was likely. Representations made beforehand can be discovered and could destroy your credibility and your client's case.

6.3 PRICING ENGINEERING SERVICES AND PAYMENT ARRANGEMENTS

The most common fee arrangement is a per hour fee for services. The engineering expert maintains a record of time spent and bills on an hour and fraction of an hour basis. The hourly rate may be uniform for all services: document review, investigative time, report preparation, travel, court time, etc. or it may be on a sliding scale. Research is at a lower rate, deposition and court appearances at a higher rate.

The use of such a fee schedule is at the choice of the expert and the acceptance by the client. The hourly fee set for the base rate depends on the qualifications and reputation of the expert. In a larger organization, senior people command a higher fee than junior people.

Travel costs are, in most cases, an extra charge. The custom and practice for other extra charges varies widely. For some experts, the hourly fee is all-inclusive except for travel. Other firms charge for support staff such as typists. If significant laboratory work is involved, laboratory personnel and equipment charges will be included. Camera and other photographic equipment, film processing, computer usage, and copying costs are other charges that may represent extras. Telephone charges may or may not be billed as extra charges. The use of junior level engineers and other technical support staff is a justifiable expense and is usually billed to clients.

The final question is, are you making a profit with your pricing structure? Should you charge one fee and eliminate extras? Does it pay to track telephone charges? Are your costs being properly identified and covered by the fee structure?

Another type of pricing structure is a fixed fee arrangement. This can work where the scope of the assignment can be reasonably predicted. Examples are situations where only document review is required or where a brief site inspection and report are required. Open-ended situations involving document searches, investigations, depositions, or court testimony do not lend themselves to fixed fee arrangements. As long as the scope can be adequately defined, fixed fees can be utilized. If the only cost variable is the expert's time, a small overrun in time may be worth the benefit of getting the assignment.

Fees for Forensic Engineering services should never be based on the outcome of the investigation. Contingency fees (percentage of recovery) or being paid in full when recovery is made, makes the expert an interested party. An expert cannot be viewed as giving testimony based on engineering certainty if his or her remuneration depends on how much the client recovers. The courts consider contingency fees for experts to be unethical.

A third form of fee structure is unit or task based pricing. This form of fee arrangement is used when the entire scope of service cannot be well defined initially or when the client wants to have the work performed in phases. In such cases, the project work is divided into logical tasks or phases that can be defined and scoped out. Either a fixed price or a budget is provided for each task. The authorized tasks are performed and no further work is done until the client authorization is received. When the task structure is used because the entire scope of work is not definable at the start, the information gained from the early tasks is used to refine or define the later tasks. For example, the initial tasks might consist of a meeting with client personnel, an initial walk-through of the facility, a plan review, and a review of available documentation. From these tasks, an analytical procedure could be developed for reviews of additional documents and testing of critical components. A final set of tasks could be development of remedial repairs, cost estimating, report preparation, and assistance in preparation for arbitration. At the start, the appropriate analysis could not be defined. Key components were unknown and the likely conclusions could not be identified. Only after the investigation was begun, could further tasks be defined.

In the situation where a client wants the work in phases, the driving reason may be cash flow or may be the presence of ongoing settlement negotiations. The client hopes that settlement might be achieved before undergoing the expense of a complete investigation. Perhaps settlement with one of several parties may be reached and detailed investigation into one area of defects may be avoided.

In selecting or recommending the fee arrangement for an assignment, practical considerations of budget and time constraints come in to play. You may be called in immediately after a major loss has occurred such as an industrial explosion or the partial collapse of a major facility. Your role may involve one or several of the following requirements:

- Evaluation of the extent of damage

- Design of immediate measures to stabilize the structure

- Design of temporary access structures to permit investigators to safely traverse the area

- Design of methods to physically and functionally isolate the area

- Monitoring of the temporary construction

- Preparation and performance of a causation investigation

- Preparation of repair schemes and procedures

Under such conditions there will be a great urgency to prepare the site for causation investigation and preparation of a budget beforehand is unrealistic. Time will be the driving factor and, within reason, cost a second consideration. Maintaining safety for workers and investigators will be the primary consideration. Any designs of temporary measures would favor ruggedness to handle the unknowns and ease of construction to speed the work. Appropriate OSHA and industry requirements would still have to be maintained. This is a challenging assignment that requires considerable attention to details. Such work would normally be done on an hourly basis. Long hours and considerable overtime will be involved.

When your assignment is to investigate a loss that occurred some time in the past, the time constraint may not be the same. Unfortunately, many clients do not discover that they require Forensic Engineering services until very shortly before a trial or arbitration hearing is at hand. In that case, time becomes critical though the cost of the services remains a consideration. You must provide a scope of service that can meet the time constraints, if at all possible. Sometimes it is too late for tests or simulations or obtaining reports and analyses that have a long lead-time. This must be revealed to your client, along with the effect the lack of these materials might have on the quality of the investigation.

The fee structure can reflect the value of the matter being investigated. It would not be reasonable to conduct a $10,000 investigation into a $2000 property loss. One of the most difficult assignments to budget is a small loss where a thorough investigation is not financially practical. When requested to prepare a budget, it is fair to review the scope of the potential investigation along with the amount of money in dispute. Can a fair but not exhaustive investigation be conducted for 5% or 10% of the loss or disputed sum? The ideal analysis can be predicated, but adjusting the budget to reflect the claim is good business practice. The client must be informed if you feel you are eliminating desirable, but not vital, steps in the process. The client might well be willing to pay if you believe the cost is justified. It may turn out, however, that you cannot perform a proper investigation for the sum the client feels the case justifies. In that case, you should be willing to decline the work.

A discussion of business considerations would not be complete without including the subject of billings and collections. For many professionals this is as difficult an area as cold calls. When you prepare a proposal for a potential client you should include your billing structure, as discussed earlier, as well as your billing procedure. Do you require a retainer? Do you bill monthly and require payment within 10 days? Do you charge interest on overdue bills? For small cases, do you bill only upon submission of the final report? Does the client have the leverage to dictate terms and conditions for payment? If you are retained by governmental agencies or large corporations, your billing practices will have to coincide with their requirements.

What do you do about a "problem client"? This would be one whose payments run past the due date and who has continuing work. Do you stop work? Do you contact a collection agency? The most difficult situations involve clients in trial or arbitration who are running behind in payments. The client's attorney may also have the same collection problems. If you stop work you could jeopardize the case and your relationship with the attorney. It is the attorney who is the more likely source of repeat business. Your leverage for payment may also be greatest at this time.

There is no simple solution. You have to assess the likelihood of collection in the future. Are you spending time without any possibility of catching up. Is this a client for whom you have worked before and who has eventually paid you in full? A frank discussion with the client would be the best starting point.

6.4 ASSIGNMENTS FROM CLIENTS, INSURANCE COMPANIES & ATTORNEYS

Assignments that are conducted by the Forensic engineer for insurance companies, attorneys, and other clients include the following:

- Investigative Assignments

- Directed and/or Limited Investigative Assignments

- Legal/Liability Evaluations for Attorneys

- Affidavits and Expert Reports

- Code Evaluations

Each of these has unique requirements with respect to marketing, determination of work scope and budget, methods of conducting, coordinating, and managing the project, and the expected deliverables of the assignment. Marketing, and issues related to scope and budget are discussed in other sections of this chapter. Emphasis here is given to business considerations, particularly those concerned with project management functions of communication, quality assurance, record keeping, and confidentiality issues. Thus, the following discussion will concentrate on the particular issues surrounding conduct, coordination, and general management of these types of forensic assignments.

6.4.1 Investigative Assignments

As discussed in Chapter 3, investigative assignments are commonly performed by Forensic engineers, and may include field investigations, office investigations, laboratory testing, or simply the review and analysis of data. The typical deliverable product of an investigative assignment is a report that reviews the methods of investigation, and presents the Forensic engineer's interpretations of results and conclusions regarding likely failure modes. The report may be followed by delivery of depositions, expert testimony in court, or appearances by the Forensic engineer in other legal proceedings such as mediation hearings. Investigative assignments may be complex, involving multiple experts and attorneys, or they may be quite simple in scope.

6.4.1.1 Conducting a Technical Investigation

Specific guidelines for conducting investigations are presented in Chapter 3 of this set of guidelines, and will not be discussed further.

6.4.1.2 Coordinating a Technical Investigation

The key aspect of the successful coordination of an investigative assignment is clear communication between the various involved parties, which may include the client, attorneys, other experts (parallel or sub-experts), and laboratories and testing facilities.

It is the responsibility of the lead investigator or project manager, in conjunction with the client, to establish communication guidelines for the project. Within a given Forensic Engineering firm or testing lab, it is good practice to appoint a lead investigator or project manager. Communications with the client and other involved parties should be handled by this individual.

An important tool to facilitate successful communication during large projects is a project handbook or guide that is written by the lead investigator or project manager. This handbook would include a list of involved parties, diagram of project management structure, rules or guidelines for contacting other parties, specific review and quality assurance requirements, subcontracts, a schedule with deadlines and milestones, deliverables expected, and specific task assignments. The handbook can be written for internal use within a particular Forensic Engineering firm or testing laboratory or for distribution to all involved parties. In the absence of such a document, the project coordinator should clearly establish communication guidelines either internally, or to all involved parties by other means, such as letters, memos, or by speaking directly to the various parties.

Specific communication restrictions should be clarified with the client at the outset of the project and documented in the project handbook or communicated to project participants by other means. For instance, the client may want to maintain the independence of various experts involved in the project. In such a case, communications between experts would be conducted through the client or an attorney, or may be limited to direct spoken communication. In other instances, the client may want to limit internal or external written communication. In either case, the project manager should establish these principles by direct spoken communication with all project participants.

Communications between experts and attorneys should be conducted or coordinated by the lead investigator within a given Forensic Engineering firm. Communications with attorneys may be protected by attorney-client privilege. If maintaining privilege is important to a project, this fact should be established early in the project and communicated to involved parties. Written material should be thoroughly reviewed and vetted by the project manager before it is sent to an attorney. Working drafts should be destroyed because they are discoverable. Each page of written communication between experts and attorney should be clearly labeled with, *"Confidential and Privileged - Attorney-Client Work Product."*

The maintenance of confidentiality is usually important to forensic investigations. Documents should be controlled and protected. It is good practice to use file folders that are color coded to confidential projects and the files should be stored in locking file cabinets. Access to these files should be limited to those directly involved with the project that have a need to use the documents.

Documents should be cataloged using a database or similar means. When received, each document should be given a unique number, such as a Bates Number, and filed by number. The document database should include the document number, author, document date, document source, date received, file location, and a brief description of the contents.

6.4.2 Directed and/or Limited Investigation Assignments

Directed and/or limited assignments are defined by a specific scope and may be a small component or a sub-expert role of a larger investigation. These assignments generally follow the same guidelines as investigative assignments with respect to the project functions of communication, quality assurance, record keeping, and confidentiality. When used as a sub-expert, the Forensic engineer should closely follow specific guidelines established by the lead investigator or client. If no such guidelines are provided, it is up to the practitioner to clarify these policies with his or her client.

6.4.3 Reporting of Bad News

As discussed previously, a Forensic engineer should not advocate for his or her client. The investigation and evaluation performed by the Forensic engineer should be consistent with sound engineering and scientific principles. There will be instances where the investigator will determine that the party, on whose behalf the Forensic engineer is engaged, made an error or acted so as to contribute to the causation of the defect or failure. There may be instances where an error was made which was not a causation but which will likely be discovered by other parties and used against the client.

Reputable attorneys and insurance adjusters want to know how strong their case is and if indeed, their principal is culpable. Reporting "bad news" is as much a part of the Forensic engineer's work as reporting "good news" which would exonerate the principal. In both instances the investigation, analysis and evaluation must be done in the most careful manner so as not to mislead the client into believing his or her position is either correct or is indefensible.

Bad news must be reported as clearly as good news. How bad is bad? Was the designer flagrantly careless in performing the design calculations? Do the design details go against all sound principals? Did the designer make a mistake in a mathematical calculation or misinterpret a detail? Designers make errors, Errors and omissions insurance is sold to cover errors. Flagrant mistakes may raise the possibility of higher jury awards against the principal. The details and significance of the errors should be carefully presented so that the attorney or adjuster can view the results in light of possible defenses, policy coverage and settlement strategies. A bad position will motivate your client to settle rather than continuing on through the litigation process.

Bad news should never be reported in writing, If your client is an attorney, he or she does not have to disclose your results if you do not become a testifying expert. If you were not revealed as an expert prior to your concluding the investigation, then your results are privileged and not discoverable. If your client is an insurance adjuster, your findings may not be privileged. Not having a written report reduces the chances of other parties discovering your results. Your client will decide if he or she wants a written report.

What if you find that there is an imminent danger to the public welfare? For instance, you may discover that a structural member is under-designed and could collapse at any time. You may go to a. house to investigate a claim about a bulging foundation wall and find that the front wall of the basement has bowed out over a foot, and earth and water are coming through the gaps in the concrete blocks. Should you make your results public? This is a difficult decision, which can only be handled on a case by case basis. If there is cause for urgent action, your client should be made aware of the urgency and appropriate actions suggested. If no actions are taken in a reasonable amount of time, then further action on your part might be warranted. In the case where the foundation wall was truly in imminent danger or collapse, the author went to the Building Official of the municipality in which the house was located and informed him of the condition. It was suggested by the author that the house either be vacated or the structure shored against collapse. The Building

Official sent an inspector to the house. The author was satisfied that the matter was placed in the hands of the appropriate public official.

6.5 INSURANCE FOR FORENSIC ENGINEERING

6.5.1 Introduction

There can be no doubt that litigation and insurance are integral aspects of the modern lifestyle. Insurance policies are readily available to cover almost every aspect of our lives including our property, our actions and even our future. The cost of insurance policies are driven by a number of factors including the coverage provided, the global history of risk associated with the coverage and the normal pressures of the marketplace. Although a policy is written to provide coverage for and protect the financial well being of the named insured, those making valid claims benefit from well written and managed policies through the protection provided the claimant against the negligence of the insured. One of the best examples is the automobile policy required by most states. It only takes one accident with an uninsured or under insured motorist to understand the importance of adequate protection.

Insurance is also an integral aspect of the engineer's daily practice which can affect the type of services offered and the limit the engineer's sphere of potential clients. In some cases, the fees generated by a firm for a particular service may not justify the cost of the insurance premiums. In other cases, the client may require the consultant maintain levels of insurance in excess of what the consultant normally carries for years beyond the completion of the project. Clients frequently demand contract language that effectively nullifies coverage while at the same time requiring the coverage to be in place. Hopefully, insurance issues are only one of many considerations made when determining what type of services a consultant provides and to whom those services are provided. However, the issue of litigation and insurance cannot be ignored by consultants providing Forensic Engineering services.

The legal aspects of the modern business world have had a profound impact on the business of engineering. Unfortunately, few engineering schools prepare graduating engineering students for what is typically their first assignment, the construction site, much less the implications of carelessly worded revisions to the project specifications. Terminology has changed in the industry to reflect a more accurate description of the roles of all parties in the field of engineering and to more accurately define responsibilities in legal terms. For example a "cost estimate" is now commonly described as an "opinion of probable cost". The "construction inspector" is more commonly referred to as the resident project representative. When reviewed by the consultant, shop drawings are no longer stamped "approved" or "rejected", they are stamped "no exceptions taken" or "resubmit". These terms are commonly defined in the construction documents. Consultants and clients alike casually describe some terms or phrases in standard contract documents or reports as "weasel words" when a better description could clarify meaning.. Clauses limiting liability, limiting the use or distribution of documents and hold harmless clauses are more and more commonplace.

6.5.2 Types of Insurance

No one insurance policy is sufficient to cover all aspects of a business based engineering practice. Most businesses provide benefits in the form of workers' compensation, health, life and disability insurance for employees. Other policies may be warranted or required under different circumstances. Most forms of insurance fall under one of the following categories.

Auto liability -	Provides coverage for losses caused by injuries or damage to property and legal liability imposed on the insured for such injury or for damage to property caused in conjunction with operation of a vehicle.
Builder's Risk -	Normally carried by a contractor or owner. Also termed All Risk or Course of Construction; intended to cover both the labor and materials necessary to rebuild in the event of damage or destruction during the course of construction.
General Liability -	Provides coverage for legal liability to third parties that are not a result of a design professional's professional acts, errors or omissions. Commonly applies to exposure created by the premises and a firm's operations.
Non-Owned Auto Insurance -	Provides coverage for liability and property damage claims which may be caused by automobiles not owned or hired by the firm.
Professional Liability Insurance -	Formerly called errors and omissions insurance. Protects the design professional against claims arising from negligent acts or omissions in the performance of professional services.
Project Insurance -	Provides project-specific professional liability coverage for one or all members of a design team on a given project.
Valuable Papers Coverage -	Covers the cost of replacing or redrawing documents lost due to fire, water damage or other specified perils.
Workers Compensation -	Required by state law for all employers. Provides benefits for medical costs and lost wages caused by work-connected illness or injury.

6.5.3 Professional Liability Policies

As noted above, professional liability insurance (PLI) protects the professional against claims arising from negligent errors, acts or omissions in the performance of professional services. PLI is a claims-made policy which means that coverage is based on when the claim is reported, not when the negligent act, error or omission occurred. For example, a civil engineer designs a culvert in 1990, but it is installed in 1991. The consultant changes PLI carriers in 1992. A tropical storm occurs in November 1995 causing flooding upstream of the culvert and a claim is filed in January 1996. Hopefully, the consultant negotiated a retroactive policy and the policy in effect in 1996 covers the claim. However, policies vary and the definition of what constitutes a claim also varies.

Claims-made policies sometimes have a restriction referred to as "prior acts" that are typically tied to a retroactive date. This exclusion eliminates coverage for claims that are reported during the policy period, which arise out of negligent acts, errors or omissions that occurred before the retroactive date. The retroactive date may be the date the policy was initiated. When completing applications for PLI principals in the firm are generally asked about prior knowledge of potential claims. Any potential claims of which there is prior knowledge would likely be excluded from the policy.

Another aspect of claims-made policies is the increase in premiums that normally occur during the first few years of coverage. The coverage in the first year is typically limited to the claims that become known and are related to acts, errors and omissions that occur in the policy period. The exposure increases from year to year, therefore, the premiums increase as well. Under the guidelines established for some policies, the retroactive date is extended retroactively to the date the engineering firm was established or to the date the individual began their practice if after a period of time no claims are made.

There are many aspects of PLI policies that should be considered when evaluating policies including;

* the limits of the policy per claim and per year,

* amounts of deductibles,

* the aspects of the professional practice that are or are not covered,

* how legal expenses are applied against the coverage limit,

* what impact on coverage or exposure refusal to accept a settlement offer reached between the claimant and the carrier might have,

* who selects counsel, and

* what affect claims or the lack thereof have on premiums.

6.5.4 Liability

According to Webster's Dictionary, liability is *the state or quality of being liable* and liable is defined as *legally responsible.* We have previously defined PLI as a policy that "*protects the design professional against claims arising from negligent errors, acts or omissions in the performance of professional services*". Design is not the standard to which liability is attached. Engineering professionals provide many services that require no design including traffic impact analysis, environmental assessments, bridge inspections, hydraulic analyses, pre-purchase condition assessments, construction administration, and Forensic Engineering and investigations. However, in each of these cases, the consultant develops an opinion upon which the client, another engineer, a developer, a public agency or an attorney, will rely.

Several areas of potential liability exist within the sphere of pure Forensic Engineering including; loss of or damage to material evidence, perceived or real conflicts of interest, errors in documentation, errors in data collection and errors in the analysis of the data. These are generally considered pre-trial litigation support services. A more interesting question is whether the consultant could be sued by the client or a third party for the quality of his or her testimony or the impact of said testimony on an unrelated issue. Decisions have been rendered concerning consultants providing litigation support. Two such examples follow.

Issues relating to pre-trial support are discussed the case of American v. Matthews (Case No. 841 S.W.2d 671; 1992 Mo. LEXIS 130). In this case, American retained Matthews as professional engineers, to prepare claims for additional compensation from Zurn, the company American subcontracted with. Matthews also presented the claims at an arbitration hearing. American's claim was for $4,888,390. The arbitrators awarded $1,118,606. American filed suit against Matthews claiming negligence in its performance of professional services caused American to be unable to support their claims for additional compensation. In response to these claims, Matthews asserted the defense of witness immunity. Missouri law generally extends immunity to defamation actions against adverse witnesses. The court ruled that Matthews was not an independent fact or opinion witness and that Matthews voluntarily agreed to provide these services for a fee. As a result, Matthews agreed to assume the duty of care of a skillful professional in exchange for a fee of $350,000. The court ruled that witness immunity could not be used to bar the suit.

In most cases, a third party is affected by the opinions developed by consultants providing Forensic Engineering services. Volume 12 number 2 of the Journal of the National Academy of Forensic engineers includes "A Case Study in Forensic Engineering Liability" in which an engineer describes the experience of being sued as a result of a forensic investigation performed on an attorney's home. In this case, two independent consultants were hired by an insurance company to determine the extent of damage caused to a residence by a lightening strike. Both engineers, working without knowledge of each other's activities reached the same conclusions. The claimant was dissatisfied with the settlement and filed a suit alleging the two consultants conspired with the insurer to deny the plaintiff's rights pursuant to the insurance policy. The consultants were ultimately granted a Motion for Summary

Judgment and released from the case but only after countless hours of trial preparation and $40,000 in out-of-pocket expenses.

Failure to meet the standard of care of a professional engineer in performing and investigation can lead to problems for the client as well as the Forensic engineer. In Nicolas versus State Farm, State Farm issued a homeowner's policy that excluded loss due to foundation settlement but did cover losses resulting from leaking plumbing. An insured submitted a claim contending foundation settlement was caused by a leaking plumbing system. State Farm hired HAAG Engineering, a consultant they frequently used in cases involving foundation settlement, to investigate the claim. HAAG concluded there was no relationship between the leaky plumbing and foundation settlement without locating the leak, determining its severity or collecting soil samples. The insured sued for breach of contract and common law bad faith. The jury found in favor of the insured on both counts. On review, the appellate court held that State Farm's blind reliance on HAAG's conclusion was suspect since State Farm was aware that the consultant usually came to that conclusion in similar cases. The court also questioned HAAG's failure to locate the leak, determine its severity or take soil samples in reaching its conclusion.

6.6 SCOPE OF FORENSIC ENGINEERING SERVICES

The scope of a forensic project generally includes case review and contract negotiations, investigation or data collection, research and analysis, reporting of findings (verbal and/or written), mediation/arbitration, depositions and interrogatories, trial and project closeout. Individual activities are very different from the typical design project, but the scope of work is similar to that of a non-Forensic Engineering study. The scope of a design project generally includes scope and contract negotiation, project issues identification (regulatory issues such as environmental, traffic and permitting; or market related issues including financing or end user needs), preliminary and final design, contractor selection, permitting, construction, warranty, and project closeout. The scope of an engineering study includes project review and contract negotiations, investigation or data collection, research and analysis, reporting of findings (verbal and/or written), presentation of findings to a public agency, presentation of findings in a public forum and project closeout. Many of the phases of a design or forensic project seem to parallel one another. In fact, many non-Forensic Engineering studies involve the same or similar elements found in the forensic scope. However, the approach to the similar elements of each project type is vastly different.

A comparison can be made by evaluating the similarities in the design of a building structure versus the analysis of potential deficiencies arising from a dispute between the owner, structural engineer and builder. During the original design, the structural engineers typically make certain assumptions that simplify the analysis. Since success is in the details, the details are then developed such that the assumptions made in the engineer's analysis are valid. For example, an engineer assumes connections in a steel frame are pinned and lateral bracing is to be provided by a combination of X-bracing and masonry in-fill shear walls. The details must make these assumptions clear, especially if the connection details are designed by the fabricator's engineer.

For illustration, assume that inconsistencies or incomplete information results in rigid connections being substituted by the fabricator. The column sizes may not be sufficient to resist the resulting moments that were not accounted for in their design. Even if they are, the additional moments applied to the footings may cause loads in excess of the capacity of the footing or the allowable soil bearing pressure recommended by the geotechnical consultant. A more stringent analysis of the structure might be necessary to evaluate the impact of the change. Since the engineer can no longer make simplifying assumptions, the analysis will likely be more time intensive and costly. In addition, the analysis and the engineer's findings are much more likely to come under the scrutiny of peer review by experts hired by the different parties involved.

6.6.1 Design Verses Forensic Analysis

Forensic Engineering investigations can become particularly risky when the professional is requested to provide repair recommendations to be used for the purpose of estimating the cost of repairs or when asked to provide opinions of probable cost. Repair designs normally require significant input and decisions by the owner or end user. Input may include the basis of the design such as special requirements not normally anticipated and the level of expectation of the owner. In the design project, the level of risk accepted by the parties is or should be incorporated into the negotiations and the contract documents. For example, some clients demand clauses in contract documents that require special riders on existing policies or project specific policies. When the clients are provided the cost of such clauses as line items in the contract, the client can then decide if the protection afforded is commensurate with the cost.

That type of one-on-one communication is generally not possible when the consultant in a forensic project is asked to recommend repairs and develop an opinion of probable costs associated with those repairs. In the case _American versus Matthews_, another case relating to cost estimates made by a consultant in a forensic case was referenced (Bruce v. Byrne-Stevens & Associates engineers, Inc.). In this case, the plaintiff sued their consultant when the cost estimate developed by the consultant was the basis of the award. It was later found that the cost of repairs was significantly higher than the settlement. In _A Case Study of Forensic Engineering Liability_ a consultant in a forensic project was sued by a third party. In any project, design or forensic, repair recommendations and cost estimates must be accompanied by appropriate clarifying or descriptive information that defines the basis of the recommendations and estimates. Appropriate research, analyses and documentation must be performed to ensure the accuracy of the information provided. Where adequate information is not available to the consultant, repair recommendations and cost estimates should not be provided.

Engineer's cost estimates are rightfully considered different from contractor's estimates. The engineer bases the estimates on the engineer's past experience with similar projects along with industry accepted resources. However, the engineer is seldom willing or able to perform the repairs. In many cases it is prudent for a separate consultant (such as an estimator or contractor) to develop opinions of probable cost independent of the engineer. However, in most cases, time and financial constraints eliminate this option. In any case, repair recommendations and cost estimates must be carefully developed and documented to the standard of care in the industry.

As in any other form of consulting services, time relates to costs. Most types of investigation require some minimum determinate amount of effort to reach conclusions based in fact and to present those findings. The client sometimes desires to minimize costs and that desire can influence the accuracy of the engineer's findings. The concern is not whether or not the client's position will be supported, but is rather the accuracy of the information published by the consultant and the use of that information. Poorly conceived repair recommendations based on insufficient investigation and testing are sometimes used by third parties even though statements in the engineer's report state that said recommendations are not to be used or relied upon by any party for the purposes of repair or construction.

Scope and liability are inseparable in the forensic project just as they are in the more traditional design project or non-Forensic Engineering study. Understanding the risks associated with certain elements of the forensic project should allow the engineer to balance risk with reward.

6.6.2 Record Retention

Copies of the data and analyses used by the Forensic engineer to develop opinions, as well as the final report should be kept in the engineer's files for at least 5 to 10 years. It is a good idea not to keep irrelevant data that was not used in the final report. However, data and analyses used to prove that certain failure mechanisms were not the cause of failure should always be kept on file.

While most data and analyses are by contract the property of the owner, few clients retain these records. Few courts retain exhibits or other materials beyond a few years' time. Often, exhibits used by an engineer and marked by the court will be returned to the engineer if a formal request is made. Failure to request the material may result in it being discarded by the court.

6.7 RELATION OF BUSINESS CONSIDERATIONS TO THE HYPOTHETICAL

Three sets of engineers were employed to investigate the failure for the three parties in the hypothetical described in Appendix A. The engineers were retained by the Owner, the Designer and the Contractor. For illustrative purposes, the means by which they were selected was different for each expert. Business aspects of this selection process are described here.

6.7.1 Selection of the State's Forensic Engineer

The Attorney General of the State became involved with the collapse because of the cost of the loss, possible personal injury lawsuits and the likelihood of legal action against the parties involved in the design and construction of the bridge. Time was of the essence in selection of the Forensic engineer since investigation and recovery of evidence would have to start as quickly as possible. It was decided to develop a "short list" of three firms who could present the credentials of their proposed investigators, approaches to the investigation, fee structure and experience. Because of the emergency nature of the situation, an open bidding scenario for consulting services would not be required.

After the collapse, the Attorney General's Office received calls and brochures from several Forensic Engineering firms offering their services in this matter. The Director of the State DOT had been contacted by a former engineering colleague who was now a consultant with a Forensic Engineering firm. It was felt by the DOT and the Attorney General that recommendations would be the best way to go in the limited time available.

Names of possible candidate firms were solicited from the State DOT and from the Attorney General's Office. The DOT provided the name of one firm that had done some smaller scale investigations for the DOT over the past few years. They were located in the State and were familiar with DOT procedures. One of the attorneys in the Attorney General's Office had formerly been in private practice and had handled construction litigation cases. She had worked with a Forensic Engineering firm on several cases. Some of the work had been of a structural or foundation nature. The firm was large and diversified and employed experts in many fields. It was nationally known. Another attorney in the Attorney General's Office had contacts with other attorneys in public and private practice. He made a series of calls and received a recommendation for a large geotechnical engineering firm that also had a Forensic Engineering practice. Though not their primary source of business, they had been involved in several collapses which had received high exposure.

Interviews and presentations were quickly arranged. Though the local firm was favored by the DOT, the attorneys felt that the courtroom experience of the two other firms was better proven. The geotechnical firm was chosen because of its current work in the field. It had never done any work for the State and could not be portrayed as being biased. A fee structure was arranged with the consultant that eliminated certain overhead items and rolled them into the hourly fees. Fees were to be different for principal investigators as opposed to staff engineers.

6.7.2 Selection of the Contractor's Forensic Engineer

The Contractor had notified its insurance carrier immediately after the Contractor heard of the collapse. The Contractor had previously been involved with small failures during construction. It was aware of the need to notify its carrier about any incident which could result in a liability claim. The carrier had an ongoing relationship with a diversified Forensic Engineering firm that had previously performed structural and foundation investigations, as well as other matters. The relationship of the Forensic Engineering firm and the carrier had started some years previously with a sales call by one of the firm's principals to one of the insurance company's Executive General Adjusters (EGA). The adjuster gave the firm one assignment, was pleased with the work and became a repeat client. The insurance company felt the work was reliable and the fees were fair. When the call came in from the Contractor and was assigned to the EGA for that region, he immediately called this firm.

6.7.3 Selection of the Design Engineer's Forensic Engineer

The Design Engineer also notified its insurance carrier when it heard of the collapse. The adjuster who was assigned the case contacted local legal counsel for the company, reasoning that there was a strong probability of legal action forthcoming. The company wanted to create the proper record to protect the interests of its insured. The attorney and the adjuster recognized the need for a Forensic engineer to deal with the forthcoming failure investigation. The attorney was experienced in construction litigation and had worked with construction failure experts both locally and nationally. He contacted a Forensic engineer from the State with whom he had worked and discussed his possible needs. The Forensic engineer had structural and foundation expertise but not seismic engineering expertise. The engineer offered to locate and interview several possible candidates for the attorney's choice. This was agreed to by the attorney. The Forensic engineer called several design firms he knew were engaged in work in high seismic zones and discussed the need for an expert in seismic engineering. He also contacted the authors of several papers in the seismic engineering field and inquired as to their availability. A professionally well respected seismic engineer who was also a faculty member at a nearby state university was selected as part of a structural and geotechnical engineering team assembled by the Forensic engineer for the attorney.

6.8 REFERENCES

ASCE, 1976. *Code of Ethics*, American Society of Civil Engineers, amended 1993 and 1996.

APPENDIX A
HYPOTHETICAL FAILURE

A three span bridge over the Muddy River has collapsed. The river is located in the mid-western section of the United States. The bridge is owned and maintained by the State Department of Transportation (DOT). Completed five years earlier, the bridge was believed to be in excellent condition.

The Bridge

The bridge itself is a three span steel structure. Two large 20' x 40' bridge piers are founded on concrete piles driven in the sand bed of the River. The bridge is oriented east to west and the flow of the river is from north to south. The main clear span is 160 feet with a series of 120-foot simple span steel girders in the center portion. The two end spans are 140 feet clear and are framed with 160-foot long cantilevered steel girders. A typical hinge connection links the girder systems together in the main span.

The 36' wide road deck is a one-way reinforced concrete slab system spanning the 12' width between girders. A safety-rail concrete barrier system is installed at the edges and in the center to separate the eastbound and westbound traffic lanes.

The Failure Scene

The most prominent feature of the collapsed bridge is the failed west bridge pier that fell to the south. It was found to be lying almost parallel to the bed of the river with part of it visible above water. Elements of the failed concrete piles can be seen extruding from the bottom of the pile cap. The river is mostly open water, but had been frozen over during part of December. The weather during the Holidays was exceptionally warm.

The east bridge pier is tilted approximately 10 degrees to the north (upstream). While all steel girders remained attached to the East pier, the center span section tore loose from the east hinge connections. The entire bridge deck and structural steel west of the eastern hinge connection lies in the water to the south of the original bridge location.

The Players

The bridge is owned by the State DOT and was designed under contract by the Design Engineer. Construction was by the Contractor.

In order to maintain privilege, the State Attorney General retains the Forensic engineer. Additionally, a Plaintiff's Attorney specializing in design and construction litigation is retained by the State.

The Design Engineer is represented by his general counsel, the Design Engineer's Attorney. On the belief that a recent seismic event (an oddity for this region) could have contributed to the failure, the Design Engineer hires a Seismic Engineering Expert.

The Contractor is also represented by general counsel, the Contractor's Lawyer. On advice, the Contractor hires the Contractor's Investigators.

The Seismic Event

Earthquake activity in this portion of the mid-west is very infrequent. One event occurred seventeen months after the bridge was opened. The earthquake (Richter magnitude 5.2) was centered approximately 45 miles north of the bridge.

Since the epicenter region is sparsely populated, the earthquake drew little interest from the public, the State or the scientific community. Little is known of the response spectra or amplification of ground motions due to the abnormal nature of the quake. No other bridges in the area experienced damage. The State instituted no additional inspection of any structures as a result of the event, and no other bridges experienced damage.

The Soil Conditions

During the project's planning stage, four soil test borings were performed by the Original Soils Engineer. Soil conditions encountered included granite rock at the abutments, alluvial sands and silts at the pier locations. The depth of alluvium was 30 to 35 feet below the river bottom prior to encountering 25 to 35 feet of residual soils and ending in granite rock. Penetration resistance values for the alluvial soils varied from 3 to 8 blows per foot. The standard penetration values in the residual soils were higher, ranging from 10 to 25 blows per foot. The borings were terminated near the granite rock / residuum interface. No coring of the rock was performed due to investigative cost restraints. The as-builts revealed that piling extended 45 ft below the stream bed.

Construction Documents and Records

The original design called for American Association of State Highway and Transportation Officials (AASHTO) Type VI concrete girders. The Contractor provided an alternative design utilizing welded steel plate girders. In the original layout, all girders were simple spans. The Contractor's alternative bid, which was approved, utilized a double cantilever framing system. All other elements of the original design remained unchanged.

A complete set of design drawings and calculations for the alternative was submitted to the State for review and approval. These documents formed an adjunct to the original contract documents. The State, the Contractor and the Design Engineer retained record sets of all submittals, shop drawings, and test reports, as well as the original construction documents including the adjunct design.

Inspections

The bridge was inspected by State DOT personnel throughout the construction process. Since the bridge opened five years before the failure, only one subsequent inspection of the superstructure had been conducted. The inspection occurred two years after opening.

The inspection revealed no sign of deterioration of any concrete work. No stress cracking or evidence of fatigue of the steel was noted. Essentially the bridge was reported to be in like-new condition. No subsurface or below water inspections were conducted.

Design Conditions

The bridge was to be designed for the normal loading usually associated with bridge construction in this part of the country. The gravity live loads were to be HS20 wheel and lane loads as established by AASHTO.

Wind loading was to be 80 miles per hour typical of the AASHTO requirements for the region. As this region is very low in seismicity, no earthquake considerations are required nor are any usually given for structures in this region. The river has a moderately low gradient and the piers had been designed for flow velocities calculated for the design flood. A scour analysis, based on the state agency's estimates of design flood rates, indicated that scour during a design flood would not exceed the safe depth for lateral support of the piles.

The bridge was to have an effective service life of 50 years. Topographic and hydrologic data for the site was available to the designers from various state agencies. No additional topographic or hydraulic studies were made by the State or the Design Engineer for the purpose of the project.

The Dam Break at the Creek

Five miles upstream from the subject site, a small tributary creek drains into the river. In the late 1950's, U. S. Army Corps of Engineers built an earthen dam on this tributary to produce a small lake for flood control and recreation purposes.

Two years after the bridge was opened, the upstream dam broke and the water flow in the creek and the river was increased significantly. Later, the State climatologist's office estimated that the water levels on the river were increased to the 75 year flood level.

Steel Alternate by The Contractor's Consultant

The original design documents issued by the State allowed a contractor steel design option. The Contractor's Consultant devised an efficient framing system that utilized welded steel plate girders. Each original concrete girder was replaced with a steel section. In order to gain additional savings, the two end span girders were designed to cantilever across the piers. The center, simple span was then reduced to

120 feet. The original design called for two 140 foot end spans and a single 160 foot clear span in the center with all being simple spans.

Since the steel alternative was 16 inches in depth less than the AASHTO girders, the concrete elevation at the top of the piers was increased to accommodate the steel. The Contractor's Consultant made no separate design checks on the piers or foundations. Their work concluded at the top of the pier. All submittals regarding the steel alternative were routed to the Design Engineer for review and approval.

The Investigation Results

Based on the configuration of the debris, the State's Forensic engineer focused on the west bridge pier. A bathymetric survey was performed and underwater photographs were taken of the failed pile bent and pile cap. The State's Forensic engineer ultimately concluded that the west bridge pier pile cap failed prematurely due to undermining of the soils below the pile cap. It is his opinion that the piles failed due to the lateral loads imparted by the river flow and ice forces.

The Contractor's investigation also focused on the west bridge pier. Soil borings were performed at the original pile cap location and the soils in the upper strata are both lower in elevation and considerably different in grain size distribution than those indicated in the original geotechnical report. Destructive testing of the bridge pier, pile cap and piles indicated that the concrete strength and reinforcing were as specified in the contract documents.

The Contractor's Investigators concluded that the foundations were constructed as specified. They are of the opinion that the lack of erosion protection for the pile caps allowed the subsurface soils to be eroded. The erosion created large, laterally unsupported lengths of the concrete piles. This combined with lateral forces due to the water flow imparted unanticipated bending moments to the piles. The Contractor also reviewed the design engineer's hydraulic and scour calculations and concluded that use of more recent data would have caused the designer to use a higher design flood rate for waterway adequacy and scour countermeasure design.

The Design Engineer's Seismic Engineering Expert took a different approach by looking at the earthquake behavior of the bridge structure. Ultimately, he focused on the non-ductile performance of the concrete piles. The Seismic Engineering Expert formed the opinion that the moderate earthquake experienced by the bridge was sufficient to damage the piles to the point that they would have cracked. This would have led to deterioration of the prestressing strands, thereby significantly reducing the bending capacity of the piles.

The Legal Positions

Lawyers for the State will assert negligence against the Design Engineer for failing to obtain the proper hydrologic studies for the site. The State will also claim breach of contract against the Contractor for failure to construct the bridge per the contract documents.

The Contractor's lawyers will deny that the bridge, as constructed, was non-compliant with contract documents and will cross claim against the Design Engineer for failing to specify erosion control measures.

Attorneys for the Design Engineer will defend both claims on the premise that the failure was an act of God. They will assert that the Design Engineer produced design documents which met all code requirements and thereby did not breach the standard of care for the industry.

Summary

This ends the description of the hypothetical and the positions that each player plans to assert at trial. Each chapter of this manual describe categorical principles of practice, then relate those principles to this hypothetical. No resolution of the case is presented in any of the chapters, but the authors believe that the reader can visualize how the principles presented could be applied to the case.

APPENDIX B
COMPILED PRINCIPLES OF ETHICAL CONDUCT
FOR FORENSIC ENGINEERS

The list below is a compilation of ethical recommendations and requirements extracted from published codes of conduct and other sources. Only those principles considered applicable to Forensic Engineering are listed. References from which most were extracted are listed in the Bibliography for Chapter 4. The principles are categorized by 13 headings as follows:

A. *Forensic Engineers and Public Safety, Citizenship and Environmental Protection*

B. *Objectivity of Forensic Engineers*

C. *Competence of Forensic Engineers*

D. *Honesty of Forensic Engineers*

E. *Thoroughness of Investigation by Forensic Engineers*

F. *Relevance of Expert Engineers' Testimony*

G. *Compensation and Business Practices of Forensic Engineers*

H. *Conflicts of Interest in Forensic Engineering*

I. *Confidentiality of Forensic Engineering Work*

J. *Demeanor of Forensic Engineers*

K. *Forensic Engineers' Conduct Toward Other Engineers*

L. *Reporting Apparent Unethical Conduct of Other Forensic Engineers*

M. *Professionalism of Forensic Engineers*

COMPILED PRINCIPLES BY CATEGORY

A. Forensic Engineers and Public Safety, Citizenship and Environmental Protection

Engineers serving as expert witnesses shall:

1. Hold paramount the safety, health and welfare of the public in the performance of their professional duties (ASCE 1977). The public interest should be held paramount (ICED 1996).

2. Immediately forward knowledge of discovery of any unsafe condition to the proper authority.

127

3. Be committed to improving the environment to enhance the quality of life (ASCE 1977).

4. Know, respect and obey established laws, governments and government regulations.

B. Objectivity of Forensic Engineers

Engineers serving as expert witnesses shall:

1. Be objective - emphasize the object, or thing dealt with, rather than thoughts and feelings.

2. Be honest and impartial in serving with fidelity the public, their employers and clients (ASCE 1977).

3. Issue public statements only in and objective and truthful manner (ASCE 1977).

4. Serve as impartial arbiters who provide and explain information regarding technical matters.

5. Diligently avoid being persuaded by attorneys to advocate their position.

6. Be impartial.

7. Tell the whole truth.

8. Have the same substance of their opinion no matter who retained them.

9. Should not accept compensation in exchange for advocacy (ICED 1996).

10. Decline or terminate and engagement when any offer of compensation in exchange for advocacy is made (ICED 1996).

11. Never be an advocate of anything other than his own opinion (TCFE 1992).

12. Refuse to state opinions until completing all the needed investigations.

13. Avoid *strategizing* with the client's advocates.

14. Avoid giving any personal opinions.

15. Only accept engagements whose scope of work is to objectively clarify technical issues.

C. Competence of Forensic Engineers

Engineers serving as expert witnesses shall:

1. Perform services, and testify, only in areas of their competence (ASCE 1977).

2. Undertake to perform engineering assignments only when qualified by education or experience in the technical field of engineering involved (ASCE 1977).

3. Perform services only in areas of their competence, when qualified by education and experience (ICED 1996).

4. Undertake an engagement only when qualified to do so (NSFE 1988).

5. Decline the engagement when their own capabilities are insufficient (ICED 1996).

6. Express an engineering opinion only when it is founded upon a background of technical competence (ASCE 1977).

7. Not affix their signatures or seals to any engineering product dealing with subject matter in which they lack competence by virtue of education or experience (ASCE 1977).

8. Not affix their signatures or seals to any engineering product or document not reviewed or prepared under their supervisory control (ASCE 1977).

9. Know, and accurately describe the applicable standards-of-care.

10. Before commenting on whether others' procedures are in keeping with the applicable standard of care, experts should perform the reasonable inquiry needed to identify the standard of care and regulations in effect at the time and place other experts' procedures were implemented (ICED 1996).

11. Obtain and explain codes, standards and regulations affecting the matters in dispute (ICED 1988).

12. Not falsify or permit misrepresentation of their academic or professional qualifications or experience (ASCE 1977).

D. Honesty of Forensic Engineers

Engineers serving as expert witnesses shall:

1. Not knowingly engage in business or professional practices of a fraudulent, dishonest or unethical nature (ASCE 1977).

2. Be truthful in professional reports, statements, or testimony.

3. Avoid erroneous or exaggerated claims regarding the applicable professional engineering standards-of-care.

4. Not attempt to conceal possible oversights or errors in their own work (ASCE

1927).

5. Concede their errors when discovered, and agree with points made through cross-examination when they have been fairly made.

6. Avoid false, misleading, or inflammatory comments (ICED 1996).

7. Admit and accept their own errors when proven wrong (NSPE 1985).

8. Refrain from distorting or altering the facts in an attempt to justify their decisions (NSPE 1985).

9. Assure that graphic representations, including models and other media, are factual in nature and avoid oversimplification and misleading exaggeration (ICED 1996).

10. Avoid omitting a material fact necessary to keep statements from being misleading (NSPE 1985).

11. Avoid the use of statements intended or likely to create an unjustified expectation (NSPE 1985).

12. Separate with labels or other techniques any opinions from facts on graphic representations (ICED 1996).

13. Concede indisputable facts even when they are adverse to the client.

14. Refrain from hiding, distorting or altering facts.

15. Advise their employers or clients when their studies show that a project will not be successful (ASCE 1977).

E. Thoroughness of Investigation by Forensic Engineers

Engineers serving as expert witnesses shall:

1. Refuse or terminate engagements that do not allow them to perform the investigations needed to establish an opinion with a reasonable degree of certainty (ICED 1988).

2. Visit the site of the event involved and consider information obtained from witnesses (ICED 1988).

3. Obtain and review documents, photographs, models, maps, and other materials relating to an issue before offering comment (ICED 1996).

4. Avoid assumptions whenever possible (ICED 1996).

5. Avoid giving opinions if the investigation was not sufficient, for whatever reason, to establish a reasonable degree of certainty.

6. Consult other experts' published works and, when appropriate and possible, speak with them directly (ICED 1996).

7. Obtain information relative to the events in question in order to minimize reliance on assumptions (ICED 1988).

8. Inform their clients of the tests, investigations, or other research they need to conduct in order to formulate their opinions (ICED 1996).

9. Speak with the authors of other documents, when doing so is appropriate and possible (ICED 1996).

10. Decline or terminate an engagement when they are denied the ability to conduct all the research they believe is necessary (ICED 1996).

11. Promptly request additional information when it is needed to clarify and inform (ICED 1996).

12. Evaluate different explanations of causes and effects (ICED 1988).

13. Not limit their inquiry for the purpose of proving the contentions advanced by those who have retained them (ICED 1988).

14. Express an opinion only when it is founded upon adequate knowledge of the facts and upon an honest conviction (ASCE 1977).

15. Conduct tests and investigations personally or direct their performance through qualified individuals who should be capable of serving as expert or factual witnesses (ICED 1988).

16. Include all relevant and pertinent information in reports, statements, or testimony (ASCE 1977).

17. Not omit material facts.

18. Provide rational explanations.

19. Respect and carefully consider the opposing point of view.

F. Relevance of Expert Engineers' Testimony

Engineers serving as expert witnesses shall:

1. Include all relevant and pertinent information in professional reports, statements, or testimony (ASCE 1977).

2. Testify about professional standards of care only with knowledge of those standards that prevailed at the time in question (ICED 1988).

3. Avoid irrelevant testimony or statements.

4. Identify standards of care independent of their own preferences (ICED 1988).

5. Not apply present standards of care to past events (ICED 1988).

6. Avoid the use of statements containing a material misrepresentation of fact (NSPE 1985).

7. Avoid over-statements of the applicable standards-of-care.

8. Avoid the use of statements containing prediction of future (NSPE 1985).

9. Avoid claims that other procedures should have been used without providing the result of applying the procedure.

10. Avoid statements containing unreasonable predictions.

11. Use only illustrative devices that demonstrate relevant principles without bias (ICED 1988).

12. Not participate in untrue, unfair or exaggerated statements regarding engineering (ASCE 1977).

13. Be responsive to questioning by both sides.

G. Compensation and Business Practices of Forensic Engineers

Engineers serving as expert witnesses shall:

1. Be compensated for the services they render, irrespective of their opinions or the outcome of the issue (ICED 1996).

2. Act in professional matters for each employer or client as faithful agents or trustees.

3. Advertise professional services in a way that does not contain self-laudatory or misleading language (ASCE 1977).

4. Not accept an expert testimony engagement on any basis of contingency.

5. Not request, propose or accept professional commissions on a contingent basis in circumstances (ASCE 1977).

6. Should not accept compensation in exchange for advocacy (ICED 1996).

7. Not accept compensation from more than one party for services on or pertaining to the same project unless agreed to by all parties (ASCE 1977).

8. Be above suspicion in financial undertakings related to the case.

9. With the concurrence of the participants involved, an expert may be retained and paid by two or more of the parties interested in the issue (ICED 1996).

10. Issue no statements, criticisms, or arguments on engineering matters that are inspired or paid for by interested parties unless they indicate on whose behalf the statements are made (ASCE 1977).

11. Not offer any gift in order to secure Forensic work.

12. Be entrusted with financial undertakings in which his honesty of purpose must be above suspicion (ASCE 1927).

13. Not use confidential information as a means of making personal profit (ASCE 1977).

14. Not compete unfairly with others (ASCE 1977).

15. Avoid improper solicitation of professional employment.

16. Avoid deception in solicitation of work (NSPE 1985).

17. Maintain custody and control over materials entrusted to the expert (ICED 1988).

H. Conflicts of Interest in Forensic Engineering

Engineers serving as expert witnesses shall:

1. Refuse or terminate involvement in an engagement when fee is used in an attempt to compromise the expert's judgment (ICED 1988).

2. Avoid conflicts of interest (ASCE 1977) and the appearance of conflicts of interest (ICED 1988).

3. Determine if they or any of their associates have or ever had a relationship with any of the organizations or individuals involved and reveal any such relationships to their client (ICED 1996).

4. Promptly inform their employers or clients of any business association, interests, or circumstances that could influence their judgment or the quality of their services (ASCE 1977).

5. Reveal any past relationships with any of the parties to their clients to permit them to determine if the relationship is a conflict of interest or has the appearance of a conflict (ICED 1996).

6. Identify whether they or any of their associates have or ever had a relationship

with any of the organizations or individuals involved (NSFE 1988).

7. With the concurrence of the participants involved, an expert may be retained and paid by two or more of the parties interested in the issue (ICED 1996).

8. Diligently avoid an engagement in which their fee is connected to the outcome.

9. Not give, solicit or receive either directly or indirectly any commission, political contribution, gift or other consideration in order to secure work (ASCE 1977).

I. Confidentiality of Forensic Engineering Work

Engineers serving as expert witnesses shall:

1. Respect confidentiality about all matters discussed by and between experts, their clients and/or clients' attorneys (ICED 1988).

2. Not reveal facts, data or information obtained in a professional capacity without the prior consent of the client or employer except as authorized or required by law or professional Codes of Conduct (NSPE 1985).

3. Not disclose confidential information concerning the business affairs or technical processes of any present or former client or employer without his consent (NSPE 1985).

4. Shall not, without consent of all interested parties, participate in or represent an adversary interest in connection with a specific project or proceeding in which the engineer has gained particular specialized knowledge on behalf of a former client or employer (NSPE 1985).

5. Supply information to the news media only when authorized to do so (ICED 1996).

J. Demeanor of Forensic Engineers

Engineers serving as expert witnesses shall:

1. Carry a professional demeanor (ICED 1988).

2. Be dignified and modest in explaining their work and merit (ASCE 1977).

3. Avoid the use of showmanship, puffery or self-laudation (NSPE 1985).

4. Avoid the use of slogans, jingles or sensational language or format (NSPE 1985).

5. Avoid anger, combativeness and acrimony.

6. Be dispassionate at all times, particularly when rendering testimony or during

cross-examination (ICED 1988).

7. Be dispassionate when discussing losses or suffering.

8. Eliminate sensational, exaggerated, or disparaging statements from their opinions (ICED 1996).

9. Refrain from appearing to be part of a contest between themselves and the litigants or some other party (ICED 1988).

10. Show respect for all persons, including the opposing team and court officials.

11. Should maintain Professional demeanor (ICED 1996).

K. Forensic Engineers' Conduct toward Other Engineers

Engineers serving as expert witnesses shall:

1. Be prepared to explain to the trier of fact the differences that exist with the opinion of other professionals and why a particular opinion should prevail (ICED 1988).

2. Refuse to accept any engagement to review the work of a fellow engineer, except with his knowledge (ASCE 1927).

3. Not review the work of another engineer for the same client without their knowledge or unless the connection of such engineer with the work has been terminated (NSPE 1985).

4. Not make requests for additional information to delay proceedings or tacitly disparage the value of others' research or findings (ICED 1996).

5. Avoid statements of opinion on the *quality* [emphasis added] of another engineer's services (NSPE 1985).

6. Cooperate and communicate with other experts whenever appropriate (ICED 1996).

7. Not indiscriminately *criticize* [emphasis added] another engineer's work (ASCE 1977).

8. When experts' opinions differ, the experts involved should explain the nature of their differences and the bases of their disagreements (ICED 1996).

9. When called upon to review others' work or provide testimony in public forums, experts render a public service (ICED 1996).

10. Should maintain professional demeanor and dignity during others' testimony

(ICED 1996).

L. Reporting Apparent Unethical Conduct of Other Forensic Engineers

Engineers serving as expert witnesses shall:

1. Report to the proper authorities any knowledge or belief that other engineers are guilty of unethical, negligent or illegal practices (NSPE 1985).

2. Notify the proper authority in writing if they have knowledge or reason to believe that another person of firm may be in violation of any of ASCE's Canons of Ethics (ASCE 1977).

M. Professionalism of Forensic Engineers

Engineers serving as expert witnesses shall:

1. Act in such a manner as to uphold and enhance the honor, integrity, and dignity of the engineering profession (ASCE 1977).

2. Recognize that they will be regarded as representatives of their profession (ICED 1996).

3. Be dignified and modest in explaining their work and merit, and avoid any act tending to promote their own interests at the expense of the integrity, honor and dignity of the profession (ASCE 1977).

4. Be an ambassador to the court for the engineering profession.

Index

examples 22; interviews 24–25; involved parties 20; photographs 23; safety 25; sampling 24; scheduling 21, 22; sketches 23; staff 21 investigations, laboratory 26–28 investigations, office 28–31; codes 30; construction documents 29–30; construction history 30; data gathering 28–31; design drawings 29; geotechnical reports 29; maintenance history 30; project files 29–30; shop drawings 31; specifications 29; structure modifications 30; test data 29

jurisdiction 11

knowledge, of engineering 3–4, 14
Kumho Tire Co. Ltd., et al., versus Patrick Charmichael, et al. 5, 95

lawyers: *see* attorneys
legal system 82–98; description 82–83; engineer as witness 84–88; engineers as consultants 96–98; expert witness qualifications 89–92; mediation 88–89; testimony admissibility 92–96
liability: and construction claims 97; of expert witnesses 50–51; insurance 116–117; standards for engineering products 46
licensing 10
litigation 36, 82

maintenance history 30
malpractice 46
marketing 99–106; niche 106; scope of 103–106
media, news 25
mediation 83, 88–89
modifications, structure 30

NAFE: *see* National Academy of Forensic Engineers
National Academy of Forensic Engineers: ethical guidelines 36;

membership 14
National Council of Examiners for Engineering and Surveying 6–8
National Society of Professional Engineers: code of ethics 54–55; ethical guidelines 36, 52; licensing requirements 6–8
NCEES: *see* National Council of Examiners for Engineering and Surveying
negligence 47, 50
Nichols, Terry 63, 65, 91
Nicolas v. State Farm 117
NSPE: *see* National Society of Professional Engineers

objectivity 3, 25, 40, 43–45, 98, 128
Oklahoma City bombing trial 63, 65, 75, 91

parties, involved 20
payments 106–109
photographs 23
presentations, to prospective clients 102–103
pricing 106–109
professional societies: Interprofessional Council on Environmental Design 15; National Academy of Forensic Engineers 14; National Council of Examiners for Engineering and Surveying 6–8; National Society of Professional Engineers 6–8
professionalism 3, 134–136
project files 28
public trust 38

qualifications: challenges to 91–92; disclosure of 91; expert 2–17, 89–92; presentation of 103

reasonable degree of certainty 49–50
reasonableness 93–94
referrals 101–102
registration, professional 8–9
remediation 67–68